T0255999

# SpringerBriefs in Applied Sciences and Technology

SpringerBriefs present concise summaries of cutting-edge research and practical applications across a wide spectrum of fields. Featuring compact volumes of 50–125 pages, the series covers a range of content from professional to academic.

Typical publications can be:

- A timely report of state-of-the art methods
- An introduction to or a manual for the application of mathematical or computer techniques
- A bridge between new research results, as published in journal articles
- A snapshot of a hot or emerging topic
- An in-depth case study
- A presentation of core concepts that students must understand in order to make independent contributions

SpringerBriefs are characterized by fast, global electronic dissemination, standard publishing contracts, standardized manuscript preparation and formatting guidelines, and expedited production schedules.

On the one hand, **SpringerBriefs in Applied Sciences and Technology** are devoted to the publication of fundamentals and applications within the different classical engineering disciplines as well as in interdisciplinary fields that recently emerged between these areas. On the other hand, as the boundary separating fundamental research and applied technology is more and more dissolving, this series is particularly open to trans-disciplinary topics between fundamental science and engineering.

Indexed by EI-Compendex, SCOPUS and Springerlink.

More information about this series at http://www.springer.com/series/8884

Hamid Reza Rezaie
Mohammadhossein Esnaashary
Abolfazl Aref arjmand
Andreas Öchsner

# A Review of Biomaterials and Their Applications in Drug Delivery

 Springer

Hamid Reza Rezaie
Department of Engineering Materials,
  Ceramic and Biomaterial Division
Iran University of Science and Technology
Tehran
Iran

Mohammadhossein Esnaashary
Department of Engineering Materials,
  Ceramic and Biomaterial Division
Iran University of Science and Technology
Tehran
Iran

Abolfazl Aref arjmand
Department of Engineering Materials,
  Ceramic and Biomaterial Division
Iran University of Science and Technology
Tehran
Iran

Andreas Öchsner
Faculty of Mechanical Engineering
Esslingen University of Applied Sciences
Esslingen
Germany

ISSN 2191-530X          ISSN 2191-5318   (electronic)
SpringerBriefs in Applied Sciences and Technology
ISBN 978-981-10-0502-2          ISBN 978-981-10-0503-9   (eBook)
https://doi.org/10.1007/978-981-10-0503-9

Library of Congress Control Number: 2018935858

Printed on acid-free paper

This Springer imprint is published by the registered company Springer Nature Singapore Pte Ltd.
The registered company address is: 152 Beach Road, #21-01/04 Gateway East, Singapore 189721, Singapore

# Contents

# Abstract

Perspective drug delivery systems can be defined as mechanisms to introduce therapeutic agents into the body to help or improve the tissue operation. Different kinds of drug delivery systems have been applied in various organs. Polymeric systems, ceramic particles, or composites in different forms of application such as injectable, coatings of implants, scaffolds, or implantable devices have been used up to now. Such a system should be able to reserve the therapeutic agent and release it at a distinct time with uniformity of the amount of agent released. With nanotechnology appearance in all parts of science, advanced drug delivery systems have been also introduced for different therapeutic applications.

**Keywords** Biomaterial · Drug delivery systems · Nanoparticle
Drug uptake and release

# Chapter 1
# The History of Drug Delivery Systems

## 1.1 Introduction

Each drug has some pharmacological properties that can produce biological effects on the human body; these effects are engendered by the interactions between the drug and specific receptors at the drug's site of action. The intensity of interactions is determined by the capability of the drug to reach its site of action. In other words, even though the medicine should be intrinsically and biologically able to cure a disease, there are some other important factors that affect the performance of medication via drugs. These factors are based on the modes in which the drugs are delivered to the site of actions (spatial placement), in addition to the rate in which the drug is released (temporal delivery) [1, 2].

For example, when a patient takes a pill, the amount of the drug in the blood rises dramatically up to its maximum point, which is possibly at noxious level, and then swiftly descends to a level which is not adequate for helping the body to overcome the disease. For maximum efficiency, it is important for drug concentration to be maintained between two boundary concentrations: the upper one is the minimal toxic concentration (MTC) and the lower one is the minimal effective concentration (MEC). Drug concentrations below the MEC and/or above the MTC would not allow the drug delivery system to be effective, safe, and have reliable application to the patient. Moreover, it can jeopardize the other tissues viability or maybe injure the targeted injured one (Fig. 1.1) [3, 4]. An ideal drug delivery system should carry out two missions well. The first one is to deliver the minimum and indeed sufficient amount of a drug to the site of action to produce the desired therapeutic response (spatial placement of the drug); and the second task is to deliver the drug with the most advantageous kinetics to amplify its beneficial responses and to downgrade undesirable side effects (temporal delivery of the drug) [2, 5].

Controlling the rate of drug release and/or the site in body where the drug is released has been attracted wide attentions and become among hot research topics

© The Author(s) 2018
H. Reza Rezaie et al., *A Review of Biomaterials and Their Applications in Drug Delivery*, SpringerBriefs in Applied Sciences and Technology,
https://doi.org/10.1007/978-981-10-0503-9_1

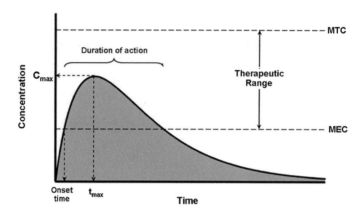

**Fig. 1.1** Typical plasma concentration-time profile after oral administration of a drug from an immediate-release dosage (Reprinted by permission from Springer Customer Service Centre GmbH [7])

for the last half-century in a multidisciplinary approach that various sciences such as chemistry, materials science, bioengineering, biology, medicine, etc. are involved [6].

Since considering the history of any subject, always equally means to ignore the growth and development of ideas that have led to the present technology, in this chapter, we briefly outline the historical evolutional achievements in the field of controlled drug delivery. Reviewing the evolution of drug delivery systems prepared our mind to deeply understand the concepts and underlying scientific principles, thus enabling us to introduce new notions in the field of controlled drug delivery.

## 1.2   Early Drug Delivery Systems

Historically, primitive ways developed by humans to enter drugs into the body were as follows: (1) chewing the leaves and roots of medicinal herbs, (2) smoke inhalation of burning medicinal materials, and (3) rudimentary juices from plants and animals. Although these seminal methods were helpful, those manners of delivering of drugs missed steadiness, uniformity, and particularity [5].

The controlled release can be traced back to over 1000 years in which pharmaceuticals were delivered through coated pills. For the first time, two Persian celebrated alchemists, Rhazes (865–925) and Avicenna (980–1037), recommended coating of pills with mucilage by applying an extract of psyllium and silvering, and gilding pills, respectively. Although their aim was to mask the taste of the bitter tasting agents, the coating still altered the release rates of the coated drugs. In the nineteenth and twentieth centuries, "Pearl coating" was another technique involving coating pills with a talc-mucilage composition that converts the pills into resembled pearls. The first so-called "sustained release" products were designed in late 1940s and early 1950s in the form of coated tablets in which the coating and the drug were layered alternatively so as to produce the ability to release the drug periodically. In this way, the period of drug action was increased, while the essential reputation of dosing decreased [8]. The coatings were non-swelling and hydrophobic at the acidity of stomach, but become ionized once the layered drug entered the slightly alkaline pH of the intestinal region of the gastrointestinal tract, and then dissolved and released the drug. Thus, the drug release was postponed from the stomach to the small intestine. The delayed release was useful to either (1) protect the stomach from the drugs that can cause gastric irritation or (2) protect the drugs that can be destroyed in the acidic atmosphere caused by many digestive enzymes in the stomach [1]. The drawback of using of these systems was their sensitivity to physiological variables like gastric emptying [8]. In addition, the pH at which the layered drug dissolved was too high to adequately dissolve it in the small intestine, so that they were not very effective [1].

## 1.3 The First-Generation Drug Delivery Systems

Till 1950, most attempts to control drug release were unsuccessful, so scientist forced to store drugs into pills or capsules that immediately released all the loaded drug as soon as contacted with water. In 1952, SmithKline Beecham developed the first commercial oral predetermined-release formulation, Spansules®, which had the capability to sustain the release kinetics of dextroamphetamine sulfate (Dexedrine®) up to 12 h. By introducing this first commercial product, the term "controlled release" went beyond a simple concept that only referred to coated tablets. Since that time, the evolutionary history of controlled drug delivery systems (DDSs) was distinguished into three time periods (three distinct generations). The first generation (1G) was started by introducing Spansule capsules which contained hundreds of micro-pellets coated with a water-soluble wax at different thicknesses, which were individual for each micro-pellet. This design created a sustained gradual release of a drug up to 12 h (patients needed to take capsules only twice a day). It was a breakthrough which improved scientists' conception of controlled drug delivery which later was advanced by introducing different drug release mechanisms (dissolution-controlled, diffusion-controlled, osmosis-controlled, and ion-exchange-controlled mechanisms). Since then, other techniques were developed to achieve extended release of a drug in the body; for example, many drugs have been reformulated in the Spansule® system, by replacing waxy coatings with more stable and reproducible synthetic polymers that were able to be gradually dissolved [1, 9]. The newly introduced systems known as controlled-release DDSs have been improving effective patient care by affording two capabilities: (1) contrary to the conventional formulations that released in a flash, controlled-release DDSs provided a sustained level of effective drug concentration in the blood; and (2) the more control on drug release kinetics resulted in alleviated side effect related concerns that obviously make new DDSs more effective [10, 11].

## 1.4   The Second-Generation Drug Delivery Systems

Although the drug delivery systems developed from 1980 to 2010, referred as the second-generation (2G), DDSs were remarkable, but their prosperity in offering clinical formulations was not very impressive [11]. At that time, it was assumed that holding a steady concentration of drug is preferable; so, the researches were focused on developing a constant rate release of drugs (zero-order DDSs). After 10 years of vast surveys, it was recognized that zero-order release kinetics is not needfully required for a DDS to be categorized as a sustained release DDS, because zero-order release cannot keep the drug concentration at a constant level alone. For example, when a drug is administrated by oral delivery system, the drug concentration depends on the drug location in the gastrointestinal tract (GI tract). It means that when the drug is located in small intestines, the drug concentration reaches maximum concentration, while when drug reaches large intestines, the drug concentration is diminished slowly. Moreover, it is not indispensable for the drug concentration to be constantly at a fixed level, rather than, as long as drug concentration is between MEC and MTC, albeit nonstationary, the drug efficacy remains the same. Although it took about a decade for this obvious verity to be identified, this discovery caused the subsequent drug delivery systems to be more flexible in their design [10].

In this era, the smart polymers and hydrogels were developed. Their added feature, which distinguishes them from others, is their sensitivity to environmental factors alteration such as pH, temperature, light, electric field, and so on. Table 1.1 briefs the drugs delivered through various stimuli-sensitive smart polymers and hydrogels developed in DDSs [12]. In addition to these advancements, 2G was also evolved using biodegradable microparticles [13], solid implants [14], and in situ gel-forming implants [15] to deliver peptides and proteins over month-long periods.

The last 10 years of the 2G was mostly focused on development of targeted nanotechnology-based DDSs to tumors and gene delivery using various biodegradable polymers in nanoparticle structures, polymeric micelles, lipids, chitosan, and dendrimers. In other words, by understanding the drug delivery release mechanisms in 1G, drug delivery vehicles were manipulated to be used as targeted vehicles in 2G, such as pH or temperature-sensitive nanoparticles. Researchers tried to manipulate the nanoparticles properties to be able to administer them directly to blood in order to deliver much more drug to its site of action. Although all nanotechnology-based drug delivery systems proved enhanced efficacy over the control of tumors decreasing in size in small animal models, very few drugs were approved by the FDA [9, 10].

**Table 1.1** Stimuli-responsive hydrogels in drug delivery (Reprinted from [12], Copyright (2002), with permission from Elsevier)

| Stimuli | Polymer | Drug |
|---|---|---|
| Magnetic field | Ethylene-co-vinyl acetate (EVAc) | Insulin |
| Ultrasonic radiation | EVAc, Ethylene-co-vinyl alcohol | Zinc bovine insulin, insulin |
| Electric field | Poly(2-hydroxyethyl methacrylate) (PHEMA) | Propranolol hydrochloride |
| Glucose | EVAc | Insulin |
| Urea | Methyl vinyl ether-co-maleic anhydride | Hydrocortisone |
| Morphine | Methyl vinyl ether-co-maleic anhydride | Naltrexone |
| Antibody | Poly(ethylene-co-vinyl acetate) | Naltrexone, ethinyl estradiol |
| PH | Chitosan–poly (ethylene oxide) (PEO) | Amoxicillin, metronidazole |
| | Poly(acrylic acid):PEO | Salicylamide, nicotinamide, clonidine hydrochloride, prednisolone |
| | Gelatin–PEO | Riboflavin |
| | PHEMA | Salicylic acid |
| | Poly(acrylamide-co-maleic acid) | Terbinafine hydrochloride |
| Temperature | Poly(N-isopropyl acrylamide) | Heparin |
| pH and temperature (Dual-Stimuli-Sensitive) | Poly(N-isopropyl acrylamide-co-butyl methacrylate-co-acrylic acid) | |

## 1.5 The Third-Generation Drug Delivery Systems

The main issue of 2G DDSs is insufficient comprehension of the side effects of DDSs on human body and the inability to overcome biological barriers by changing the physicochemical properties of DDSs. Briefly, the 2G systems feature the main point that the key to prosperity in developing beneficial DDSs is to overcome the biological barriers, in addition to physicochemical barriers, simultaneously. Hence, pharmaceutical research should gather and employ different disciplines (e.g., material science, engineering, biology, physiology, and computer science) to overcome physicochemical and biological barriers for new DDSs. Although new materials always attract lots of attention in the drug delivery research, we still need to know the exact effect of using each new material on the body's physicochemical and biological properties. The physicochemical problems originate from the poor water solubility of drugs, large molecular weight of peptide and protein drugs, and difficulty of controlling drug release kinetics. On the other hand, the biological barriers depend on overcome issues in distribution of DDSs in body and increase its half-life circulation. [9, 16]. Table 1.2 lists some of the frontier barriers for developing successful 3G DDSs and Fig. 1.2 illustrates the evolution of drug delivery systems.

Poor water solubility causes some issues such as impaired bioavailability and increased cost of drug product. For example, low absorption of the drug in the gastrointestinal tract for oral administration is happening when drugs have low solubility in water. On the other hand, low solubility may cause precipitation and aggregation of drugs that provoke some toxic effects. Therefore, great efforts have been made to improve the solubility of the drug candidates [17].

For maintaining health, there are some bioactive agents that have responsibilities in controlling performance of the body such as proteins and polypeptides. These drugs should be administrated in predetermined quantities, for targeted body site and the right time [18]. These macromolecular drugs are usually delivered by parenteral administration but their big dimensions avoid its absorption by the digestive tract. So, new methods of delivering are applied such as pulmonary, nasal, and transdermal methods [16].

As mentioned above, drug delivery is described as a system organized to release a drug at predetermined time and release rate. In designing a suitable carrier, one can face with two challenges of organizing the carrier to target a special place and release in predetermined rate.

Designing a carrier system for drug delivery applications is quite challenging in terms of targeting the drug to a specific site and continuous release over a specific period of time. In the next chapters, we introduce different drug carriers, their release mechanism, and how to decorate them with biological agent to conduct them to the predetermined site.

**Table 1.2** Barriers to overcome by the 3G drug delivery systems (Reprinted from [16], Copyright (2015), with permission from Elsevier)

| Delivery technology | Formulation barriers | Biological barriers |
|---|---|---|
| Poorly water-soluble drug delivery | New excipients for increasing drug solubility | Nontoxic to the body<br>No drug precipitation in the blood |
| Peptide/protein/ nucleic acid delivery | Control of drug release kinetics<br>Control of drug loading<br>Control of therapeutic period | IVIVC<br>Long-term delivery up to a year<br>Noninvasive delivery |
| Targeted drug delivery using nanoparticles | Control of nanoparticle size, shape, surface chemistry, functionality, and flexibility<br>Surface modification with ligands<br>Stimuli-sensitive delivery systems | Controlling biodistribution through altering vascular extravasation, renal clearance, metabolism, etc.<br>Navigating microenvironment of diseased tissues to reach target cells<br>Crossing endothelial barriers (e.g., blood–brain barrier)<br>Crossing mucosal barriers |
| Self-regulated drug delivery | Signal specificity and sensitivity<br>Fast responsive kinetics<br>Ability to stop drug release | Functional inside the body<br>Functional over the lifetime of drug delivery |

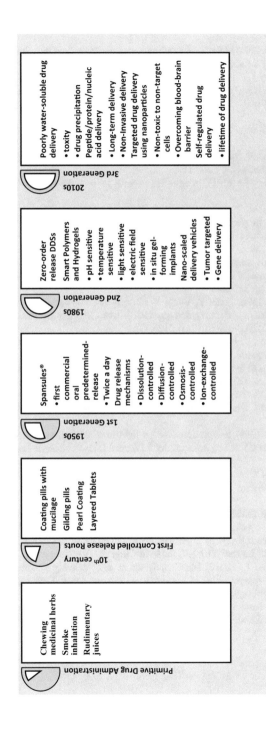

**Fig. 1.2** Chronological evolution of controlled drug delivery systems (Reprinted with modification [16], Copyright (2015), with permission from Elsevier)

# References

1. Hillery AM, Park K (2017) Drug delivery: fundamentals & applications, 2nd edn. CRC Press
2. Gennaro AR (2000) Remington: the science and practice of pharmacy, Twentieth. Lippincott Williams & Wilkins
3. Langer RS, Peppas NA (1981) Present and future applications of biomaterials in controlled drug delivery systems. Biomaterials 2:201–214. https://doi.org/10.1016/0142-9612(81) 90059-4
4. Perrie Y, Rades T (2009) Pharmaceutics: drug delivery and targeting, 1st edn. Pharmaceutical Press
5. Akala EO (2004) Oral controlled release solid dosage forms. In: Theory and practice of contemporary pharmaceutics. CRC Press
6. Wilson CG, Crowley PJ (2011) Controlled release in oral drug delivery. Springer, US, Boston, MA
7. Mehrotra N, Gupta M, Kovar A, Meibohm B (2007) The role of pharmacokinetics and pharmacodynamics in phosphodiesterase-5 inhibitor therapy. Int J Impot Res 19:253–264. https://doi.org/10.1038/sj.ijir.3901522
8. Akiti O, Jimoh AG, Wise DL, Barabino GA, Trantolo DJ, Gresser JD (1996) Multiphasic or "Pulsatile" controlled release system for the delivery of vaccines. In: Wise DL, Trantolo DJ, Altobelli DE, Yaszemski MJ, Gresser JD (eds) Human biomaterials applications. Humana Press, Totowa, NJ, pp 319–343
9. Zhang W, Zhao Q, Deng J, Hu Y, Wang Y, Ouyang D (2017) Big data analysis of global advances in pharmaceutics and drug delivery 1980–2014. Drug Discov Today 22:1201–1208. https://doi.org/10.1016/j.drudis.2017.05.012
10. Park K (2014) Controlled drug delivery systems: past forward and future back. J Control Release 190:3–8. https://doi.org/10.1016/j.jconrel.2014.03.054
11. Park K (2016) Drug delivery of the future: chasing the invisible gorilla. J Control Release 240:2–8. https://doi.org/10.1016/j.jconrel.2015.10.048
12. Gupta P, Vermani K, Garg S (2002) Hydrogels: from controlled release to pH-responsive drug delivery. Drug Discov Today 7:569–579. https://doi.org/10.1016/S1359-6446(02) 02255-9
13. Brannon-Peppas L (1995) Recent advances on the use of biodegradable microparticles and nanoparticles in controlled drug delivery. Int J Pharm 116:1–9. https://doi.org/10.1016/0378-5173(94)00324-X
14. Rosen HB, Chang J, Wnek GE, Linhardt RJ, Langer R (1983) Bioerodible polyanhydrides for controlled drug delivery. Biomaterials 4:131–133. https://doi.org/10.1016/0142-9612(83) 90054-6
15. Hatefi A, Amsden B (2002) Biodegradable injectable in situ forming drug delivery systems. J Control Release 80:9–28. https://doi.org/10.1016/S0168-3659(02)00008-1
16. Yun YH, Lee BK, Park K (2015) Controlled drug delivery: historical perspective for the next generation. J Control Release 219:2–7. https://doi.org/10.1016/j.jconrel.2015.10.005
17. He CX, He ZG, Gao JQ (2010) Microemulsions as drug delivery systems to improve the solubility and the bioavailability of poorly water-soluble drugs. Expert Opin Drug Deliv 7:445–460. https://doi.org/10.1517/17425241003596337
18. Rawat M, Singh D, Saraf S (2008) Lipid carriers: a versatile delivery vehicle for proteins and peptides. Yakugaku Zasshi 128:269–280. https://doi.org/10.1248/yakushi.128.269

# Chapter 2
# Classification of Drug Delivery Systems

## 2.1 Introduction

As stated in the last chapter, a drug delivery system is responsible to influence and determine the concentration profile of a drug, the kinetics of drug release, the site, and duration of drug action, and finally prevent the unwelcome side effects of a drug. Hence, the role of an optimal drug delivery system is to make sure that a drug is available at the site of action at specified times and within the necessary periods of time. This important role is pertained to the frequency of dosing, drug release rate, and the route of administration [1]. So, new ideas on controlling the pharmacokinetics, pharmacodynamics, nonspecific toxicity, immunogenicity, bio-recognition, and efficacy of drugs were generated. These new strategies are often called drug delivery systems (DDSs) [2]. In another word, DDSs are like a bridge between a patient and a drug. Sometimes, the formulation of a drug is administered for a remedial purpose, or maybe different mechanisms or devices are served to deliver a fixed formulation [3]. Because this book mainly focuses on finding the best method for drug delivery, it is necessary to examine the different categorizations of DDSs.

Basically, the DDSs can be divided into two main types: (1) conventional DDSs and (2) novel DDSs that are sometimes called controlled DDSs.

## 2.2 Conventional Drug Delivery Systems

Conventional DDSs are classical methods for delivery of a drug into the body. Generally, these systems are used more often when the goal is quickly absorption of a drug; therefore, a quick release of the drug is required [4]. The conventional drug delivery forms include simple oral, topical, inhaled, or injections methods [5]. These methods cannot keep the drug concentration at a fixed and constant level for a given period of time (temporal delivery). One solution to overcome the problem

© The Author(s) 2018
H. Reza Rezaie et al., *A Review of Biomaterials and Their Applications in Drug Delivery*, SpringerBriefs in Applied Sciences and Technology,
https://doi.org/10.1007/978-981-10-0503-9_2

**Fig. 2.1** Common drug
release profiles (Reprinted
from [4] with permission of
The Royal Society of
Chemistry)

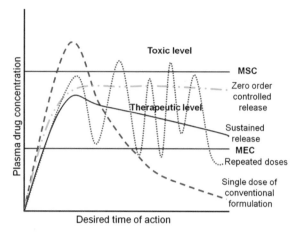

Desired time of action

of drug instability concentration is administration of multiple doses at regular intervals (repeated doses). However, this method has its own limitations. As Fig. 2.1 illustrates, the concentration of the drug varies up and down irregularly in blood plasma and patient typically forgets to take the specific dose at its exact time. Due to the problems mentioned for conventional DDSs, the necessity of providing novel DDSs becomes more apparent [4, 6].

## 2.3  Novel Drug Delivery Systems

Novel drug delivery system (NDDS) also called controlled-release drug delivery system is a combination of advanced techniques and new dosage forms to introduce better drug potency, control drug release, provide greater safety, and target a drug specifically to a desired tissue [7]. The term "controlled release" has a meaning that goes beyond the scope of only sustained release action. In other words, controlled release must have two properties such as predictability and reproducibility in the release kinetics. NDDSs lead to efficient use of expensive drugs and excipients, and reduce in production cost. From the patient point of view, NDDS brings better therapy by improved comfort drug delivery devices which increase the standard of living [8]. Table 2.1 illustrates the advantages and disadvantages of controlled-release DDSs [9, 10]

NDDSs are divided into four categories including (1) rate-preprogrammed, (2) activation-modulated, (3) feedback-regulated, and (4) site-targeting DDSs [11, 12].

**Table 2.1**  Advantages and disadvantages of controlled-release drug delivery [9, 10]

| Advantages of controlled-release routs | Disadvantages of controlled-release routs |
|---|---|
| Extension of the duration of action and bioavailability of the drug | Possibility of toxicity of the materials |
| Minimization of drug degradation and loss | Harmful degradation products |
| Reduction of dosing frequency | Necessity of surgical intervention either on systems application or removal |
| Minimization of drug concentration fluctuations in plasma level | Patients discomfort with DDS device usage (pain caused by the presence of the implant) |
| Improved drug utilization | High cost of final product due to carrier cost or fabrication procedure |
| Improved patient compliance | |
| Maintaining drug concentration in therapeutically range | |
| Reducing (eliminating) side effects by local administration | |
| Protecting short half-life drugs from degradation | |
| Improving patient compliance | |
| Less expensive and less wasteful drug delivery | |
| Facilitated administration in underprivileged areas | |

## 2.3.1   Rate-Preprogrammed Drug Delivery Systems

In this system, the rate or speed of a drug release is controlled by modification in the system design that controls the diffusion rate of drug molecules across the barrier medium surrounding it. Fick's laws of diffusion are commonly applicable for diffusion rate determination. These systems can be further classified as (1) polymer membrane permeation-controlled DDSs, (2) polymer matrix diffusion-controlled DDSs, (3) polymer (membrane/matrix) hybrid-type drug delivery systems, and (4) micro-reservoir partition-controlled DDSs [11, 12].

(1) *Polymer membrane permeation-controlled DDSs*: In this system, a drug is completely or partially loaded in a reservoir compartment (such as solid particles, a dispersion of solid particles, or a concentrated solution in a liquid- or solid-type dispersing medium). Different shapes of reservoirs can be coated with a polymeric membrane which controls the rate of drug release (Fig. 2.2). The drug release rate from this type of DDS is controlled at a preprogrammed rate by modulating the thickness of the membrane and the diffusivity of the drug [12].

(2) *Polymer matrix diffusion-controlled DDSs*: To fabricate this kind of DDS, drug particles are being dispersed homogeneously in either lipophilic (non-swellable) or a hydrophilic (swellable) polymer matrix (Fig. 2.3). The dispersion process is carried out by blending fine drug particles with either thermoset or thermoplastic polymers. The resultant drug–polymer dispersion is then molded or extruded to form drug delivery devices in various shapes and sizes. The drug release rate from this type of DDS is controlled at a preprogrammed rate by specifying the amount of drug loaded in the polymer, or adjusting diffusivity of drug inside the polymer matrix [11].

**Fig. 2.2** Various shapes of polymer membrane permeation-controlled DDSs: sphere-type (**a**), sheet-type (**b**), and cylinder-type (**c**) (Reprinted by permission from Springer Customer Service Centre GmbH [13])

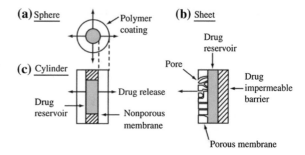

**Fig. 2.3** Drug release mechanism of a lipophilic, non-swellable polymer matrix, with a growing thickness of drug depletion zone (**a**), and a hydrophilic, swellable polymer matrix, with a growing thickness of drug-depleted gel layer (**b**) (Reprinted by permission from Springer Customer Service Centre GmbH [13])

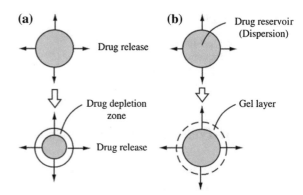

(3) *Polymer (membrane/matrix) hybrid-type DDSs*: This type of DDS is introduced in order to have both the zero-order release of polymer membrane permeation-controlled DDSs and the high mechanical properties of polymer matrix diffusion-controlled DDSs. Therefore, a non-medicated polymeric membrane is coated on surface of the drug–polymer dispersion for controlling the release of drug by membrane permeation rather than matrix diffusion [12, 13].

(4) *Micro-reservoir partition-controlled DDSs*: In this type of DDS, the drug reservoir is fabricated by high-energy microdispersion techniques in order to form a homogeneous suspension of drug solid particles in an aqueous water-miscible polymer like PEG. The resulting suspension is composed of unleachable and microscopic drug reservoirs in a biocompatible polymer which can be formed in various shapes and sizes by molding or extrusion techniques. The drug release rate from this type of DDS is controlled at a preprogrammed rate by dissolution- or matrix diffusion-controlled processes [12, 13].

## 2.3.2 Activation-Modulated Drug Delivery Systems

In this kind of DDS, the drug release is stimulated by some physical, chemical, or biochemical processes and/or external energy. So, the kinetics of release can be modulated by controlling the applying processes or the amount of external energy.

In other words, the release rate is adjusted by some devices that are able to regulate the drug release to be as much as required in physiological need. These devices are categorized as (1) externally modulated and (2) self-modulated. In externally modulated devices, an external signal is applied to the device and the device adjusts the drug release based on the intensity of the signal. On the contrary, in self-modulated, the changes in drug release are based on the environmental changing surrounding the device [14].

Activation-modulated DDSs can also be categorized based on the modulating means (Fig. 2.4) [11].

### 2.3.2.1 Activation-Modulated by Physical Means

(1) *Osmotic Pressure-Activated DDSs*: The phenomenon in which atoms or molecules spontaneously diffuse from a high concentration to low a concentration site is called osmosis. So, when drug is loaded inside a reservoir that is coated with a semipermeable membrane, water can diffuse into the reservoir to dissolve the drug (high concentration site). Due to the presence of a high concentration region inside the reservoir and a low concentration region out of it, osmotic pressure comes up and prompts the drug to go out of the reservoir through the specific hole in the coating. In this system, drug release kinetics are controlled by adjusting the diffusion rate of water and the evacuation rate of drug through the hole [15–17].

(2) *Hydrodynamic Pressure-Activated DDSs*: Like osmotic pressure, hydrodynamic pressure is also surveyed as a source of energy to modulate drug release. In this DDS system, a liquid drug is placed inside a collapsible, impermeable reservoir. Then, the reservoir is inserted inside a shape-retaining housing. An absorbent layer and a swellable hydrophilic layer also enclose the drug reservoir compartment. The laminate will imbibe the GI fluid through the annular openings at the lower end of the housing and become swollen and creates hydrodynamic pressure in the system. Then, the generated hydrodynamic pressure reduces the drug reservoir compartment volume and prompts the liquid drug to be released through the delivery orifice [11, 12].

(3) *Vapor Pressure-Activated DDSs*: In this DDS, the drug reservoir is inserted inside an infusion compartment that is separated from the pumping compartment by a movable partition. There is a vaporizable fluid (fluorocarbon, etc.) inside the pumping compartment that easily vaporizes at body temperature and creates a vapor pressure. This pressure pushes the partition to force the drug

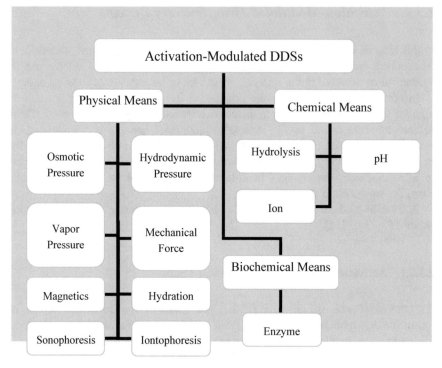

**Fig. 2.4** Classification of different activation-modulated DDSs based on the modulating means [13]

solution to be delivered through some flow regulator and delivery cannula at a constant rate [11, 12].

(4) *Mechanical Force-Activated DDSs*: In this type of DDS, the drug reservoir is equipped with a mechanically activated pumping system. The specified amount of drug is delivered into the body cavity, such as nose or mouth, through a spray system which works mechanically (pumping system). An example of this type of DDS is a metered-dose nebulizer for the intranasal administration [11, 12, 18].

(5) *Magnetic-Activated DDSs*: Many therapeutic agents like macromolecular drugs (peptides, proteins, bovine serum albumin, etc.) are known to be released very slowly from polymeric matrixes. But these macromolecular substances can be delivered with a zero-order drug release profile from a hemispheric polymeric DDS by applying magnetic energy. In some cases, electromagnetically vibration mechanism is also used. This DDS can be manufactured by placing a tiny doughnut-shaped magnet inside a drug-dispersing biocompatible polymer

matrix and coating the medicated polymer matrix with a pure polymer (ethylene–vinyl acetate copolymer or silicone elastomers). There should be a cavity at the flat surface for drug release [12].

(6) *Sonophoresis-Activated DDSs*: In this DDS, an ultrasonic device is applied for the activation of drug delivery from a polymeric device fabricated from either a nondegradable polymer (ethylene–vinyl acetate copolymer) or a bio-erodible polymer (poly [bis(p-carboxyphenoxy)alkane anhydride]) [11].

(7) *Iontophoresis-Activated DDSs*: This type of DDS uses electric current to activate and facilitate the drug delivery via penetration of charged or ionic drug molecules through the biological membrane like skin. It is very similar to passive diffusion under a concentration gradient but at a much facilitated rate. This DDS has emerged a new design of iontophoretic drug delivery system, the transdermal periodic iontherapeutic system (TPIS). TPIS has significantly improved the efficiency of transdermal delivery of peptide and protein drugs [12, 13, 19].

(8) *Hydration-Activated DDSs*: In this DDS, drug reservoirs are dispersed in a swellable hydrophilic polymer matrix (Table 2.2). The hydration-induced swelling of polymers has been used as a process to activate and modulate the release of drugs impregnated in the polymer.

#### 2.3.2.2   Activation-Modulated by Chemical Means

(1) *pH-Activated DDSs*: This type of DDS is based on the fact that pH varies from one segment of the GI tract to another. Because of irritating of drug on gastric mucosa and its instability in gastric fluid, pH-activated DDSs are mostly used to target only the intestinal tract rather than the stomach. This type of DDS is fabricated by coating a gastric fluid-sensitive drug with a combination of an intestinal fluid-insoluble polymer (such as ethyl cellulose) and an intestinal fluid-soluble polymer (like hydroxyl methylcellulose phthalate). So, the coated drugs are resistant to the gastric fluid (pH < 3) and are protected from the acidic degradation. In the small intestine, the intestinal fluid dissolves the coated

**Table 2.2** Hydrophilic polymers commonly used in hydration-activated DDSs (Reprinted from [20], Copyright (2016), with permission from Elsevier)

| Cellulosic | Non-cellulosic |
|---|---|
| Methylcellulose | Sodium alginate |
| Hydroxypropylmethylcellulose (Hypromellose, HPMC) | Xanthan gum |
| Hydroxypropylcellulose (HPC) | Carrageenan |
| Hydroxyethylcellulose (HEC) | Chitosan |
| Ethylhydroxyethylcellulose (E-HEC) | Guar gum |
| Sodium carboxymethylcellulose (Na-CMC) | Pectin |
| Methylcellulose | Cross-linked high amylose starch |

membrane of drugs due to high pH of the intestinal fluid (pH > 7.5). The rate of drug delivery can be regulated by adjusting the ratio of the intestinal fluid-soluble polymer to the intestinal fluid-insoluble polymer in the membrane composition [11, 12]. Table 2.3 names the drugs commonly delivered with different pH-sensitive smart polymers.

(2) *Ion-Activated DDSs*: This type of DDS is particularly used for controlling the delivery of ionic/ionizable drugs. Due to the constant level of ions in gastrointestinal fluid, ion-activated DDS can be modulated at a constant rate. This type of DDS is fabricated by complexing an ionic/ionisable drug with an ion-exchange resin in a granule form. For instance, complexing a cationic drug with a resin containing $SO_3^-$ groups or an anionic drug with a resin containing $N(CH_3)_3^-$ groups. Then, a water-insoluble but water-permeable polymeric membrane (such as ethyl cellulose) is coated on the granules. This membrane controls the release rate of drug from DDS. In the GI tract, hydronium and chloride ions diffuse into the drug–resin matrixes and trigger the dissociation to release ionic drug [12].

(3) *Hydrolysis-Activated DDSs*: This type of DDS is modulating the drug release profile by the hydrolysis of biodegradable/bio-erodible polymers that are saturated with drugs. The common polymers that are used in this DDS are polylactide, poly (lactide–glycolide) copolymer, poly (orthoester), and poly (anhydride), and the rate of drug delivery is controlled by polymer degradation rate.

### 2.3.2.3 Activation-Modulated by Biochemical Means

This type of DDS is based on the enzymatic hydrolysis of polymeric DDSs impregnated with drugs. In enzyme-activated DDS, the drug reservoir is either physically entrapped in microspheres or chemically bound to polymer chains fabricated from biopolymers (albumins or polypeptides). The release of drugs is triggered by the enzymatic hydrolysis of biopolymers by a specific enzyme in the targeted tissue [11, 12].

## 2.3.3 Feedback-Regulated Drug Delivery Systems

In this kind of DDS, some sensors are located on DDS devices that detects the concentration of some biochemical substances (feedback mechanisms) and the drug release is regulated by the concentration of biochemical agents. Feedback-regulated drug delivery system is divided into three categorizations as follows: (1) bio-erosion regulated DDSs, (2) bio-responsive regulated DDSs, and (3) self-regulating DDSs [11, 12].

**Table 2.3**  Various drugs delivered through pH-responsive polymers for drug delivery (Reprinted from [21], Copyright (2016), with permission from Taylor & Francis)

| Drug | Polymers | Application and Capability |
|---|---|---|
| Methylprednisolone | Carboxymethyl chitosan & Carbopol 934s | Intestinal drug delivery: Release minimum amount in acidic pH but sustained and high drug release in the basic pH of intestine |
| Clarithromycin | Chitosan and N,N'-Methylenebisacrylamide | *Helicobacter pylori* infection treatment: Drug release at lower value of pH in stomach maintained for longer period of time |
| Indomethacin | Poly (hydroxyethyl methacrylate-co-acrylic acid) | Enteric drug delivery: Swelling control mechanism is suggested for peculiar release of drug |
| Terbinafine | Poly(acrylamide/maleic acid) | *Candida albicans* infection treatment: Terbinafine adsorption capacity depends on parameters like pH of the solution and maleic acid content of polymeric hydrogel |
| Vitamin B12 & Salicylic acid | Polyvinyl alcohol | Colon-targeted drug delivery: Maximum drug release in simulated intestinal fluid where swelling pattern varied with ratio of PVA and maleic acid |
| Insulin | Poly (N-isopropylacrylamide-cobutylmethacrylate-co-acrylic acid) | Polypeptide drug carrier |
| Doxorubicin | Poly(N-isopropylacrylamide) | Anticancer activity |
| Proteins | Poly(itaconic acid-g-ethylene glycol) | Oral delivery of bioactive agents like proteins: Maintained a collapsed configuration at acidic pH, also confirmed the cytocompatibility (up to 5mg/mL), although toxicity was observed at 10mg/mL |
| Insulin | Ethylene-co-vinyl acetate (EVAc) | Sustained release of insulin from the polymer: Provide sufficient mechanical strength in physiological environment |
| Calcitonin | Poly(N-isopropyl acrylamide-co-butyl methacrylate-co-acrylic acid) | Useful for the delivery of human calcitonin: Provide maximum release of drug in less time while maintaining structural integrity |

(1) *Bio-erosion regulated DDSs*: In this DDS, the drug is loaded inside a bio-erodible polymer like polyvinyl methyl ether and then the matrix is coated with a layer of immobilized urease. In the environment with neutral pH, the polyvinyl methyl ether matrix erodes very slowly. But the urease at the surface metabolizes to form ammonia which increases the pH and enhances the degradation of the polyvinyl methyl ether matrix as well as drug release rates [11, 12].

(2) *Bio-responsive regulated DDSs*: In this DDS, a bio-responsive polymeric membrane surrounds the drug reservoir. The permeability of the membrane is dependent on the concentration of biochemical agents in the site of action [11, 12].

(3) *Self-regulating DDSs*: In this DDS, a polymeric semipermeable membrane encloses the drug reservoir. The release of drug is regulated by the membrane permeation of a biochemical agent from the tissue where the DDS is located [11, 12].

## 2.3.4   Site-Targeted Drug Delivery System

This mode is based on delivering a specific amount of a therapeutic agent for an extended period of time to a targeted diseased area of the body (such as cancerous tissues). So, the drug is only active in the targeted area in which the drug should be released. Targeted DDSs have some properties in comparison to conventional ones which attract attention of physicians. In addition to shorter half-life of drugs in conventional routs, they have poor solubility, stability, and absorption. Moreover, drugs in the conventional DDSs have low specificity and low therapeutic index as compared to the targeted DDSs [22].

Advantages:

- Reduces the side effects;
- Low toxicity;
- Avoiding the degradation of drug (first-pass metabolism);
- High drug bioavailability;
- Low fluctuation in concentration;
- Positive effect on permeability of proteins and peptides;
- Decrease dosage frequency;
- Decrease cost of expensive drug administration.

Disadvantages:

- Necessity of advanced techniques and skilled persons;
- Difficulty to maintain stability of dosage forms.

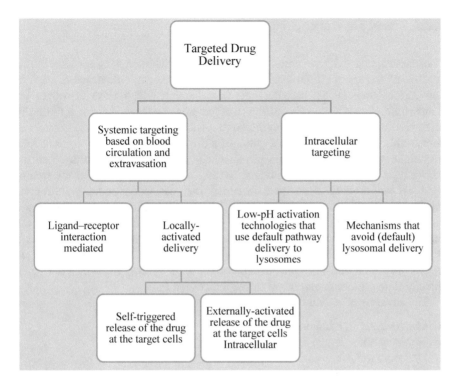

**Fig. 2.5** Classification of the targeted drug delivery processes (Reprinted with modification from [23], Copyright (2011), with permission from Elsevier)

Targeted DDSs have frequently been divided into categories of passive and active targeting.

(1) *Passive targeting*: This mechanism happens when macromolecules accumulate determinedly in the targeted tissues like tumors as a result of the enhanced permeability and retention phenomenon.
(2) *Active targeting*: This mechanism happens when specific interactions occurred between the nanocarrier and receptors on the target cell.

Figure 2.5 illustrates another categorization of targeted DDSs into two wide areas of systemic targeting and intracellular targeting [23].

## 2.4   Classification of DDSs Based on Route of Administration

Various DDSs can be categorized based on their route of administration. Route of administration is the path taken by the drug to get into the body.

## 2.4.1  Oral Delivery

The classical and vastly common type of drug administration is oral intake. In this type of administration, both local and systemic effects can be reached readily. In conventional oral DDSs, because there is no control over release of a drug (or maybe a little control), the concentration of the drug goes up and down periodically after each administration. The variation in concentration leads to side effects in some cases. Moreover, extra absorption of a drug from conventional formulations may differ greatly, depending on physicochemical properties of the drug and its carrier type, and various physiological factors such as the presence or absence of food, pH and motility of the gastrointestinal tract, etc. Uncontrolled rapid release results in increased concentration of drug further than MTC and causes some toxic side effect. Therefore, by designing the carrier to control the drug release, the mentioned problem can be dissolved. Ideal oral DDSs should steadily deliver a measurable and reproducible amount of a drug to the target site over a prolonged period [17]. Oral delivery includes tablets, capsules, syrups, etc. which are taken directly through mouth and travel through digestive tract.

The advantages of oral delivery are as follows [8, 24]:

- Convenience in administration;
- Noninvasive,
- Unit dosage form,
- Higher compliance, and
- Economical for patients.

On the other hand, oral delivery of drugs with poor aqueous solubility, low permeability and poor enzymatic and/or metabolic stability, poor gastrointestinal permeability, erratic absorption, large variations in intra- and inter-subject pharmacokinetics, and lack of dose proportionality is very challenging [8, 24].

## 2.4.2  Buccal/Sublingual Delivery

The buccal mucosa lines the inner cheek, and buccal formulations are placed in the mouth between the upper gingivae (gums) and cheek (sometimes referred to as the buccal pouch) to treat local and systemic conditions. Delivery systems in buccal delivery include mouthwashes, aerosol sprays, chewing gums, bio-adhesive tablets, gels, and patches [25, 26].

Advantages:

- Ease of administration,
- Avoidance of possible drug degradation in the gastrointestinal tract,

- By-pass first-pass metabolism,[1]
- Rapid absorption, and
- Low enzymatic activity.

Disadvantages:

- Discomfort during dissolution (mouthfeel, taste, and irritation),
- Involuntary swallowing of saliva can result in drug loss from the site of absorption,
- Short residence time, and
- Small absorption area.

### 2.4.3   Rectal Delivery

In this system, suppository is placed inside the rectum and it liquefies at body temperature to give quick effect. Since these dosage forms are neither to be swallowed nor need to be taste-masked, the rectal route is considered a good alternative for the oral administration. Due to elimination of first-pass metabolism, this form of administration is also useful for vomiting children or unconscious patients. But the main application of rectal administration is as follows: (1) for a drug that is poorly absorbed in the upper GI tract, (2) for a drug that has high first-pass effect, (3) for a drug that is unstable facing proteolytic enzymes, (4) when the presence of a drug in the gastric mucosa can cause irritation, and (5) when high doses are needed, it cannot be formulated in oral dosage [27].

On the other hand, despite to its discomfort, rectal delivery is not generally employed in situations associated with erratic absorption. So, safety, efficacy, and bioavailability evaluation of the drugs should be taken into account. It should be mentioned that using rectal administration requires great care for premature infants, because their rectal lining is very delicate and it can be easily torn and infected. For the same reason, to avoid the risk of trauma leading to possible abscess formation, rectal administration is not useful for patients having an impaired immune system [27].

---

[1]First-pass effect is a phenomenon that happens when a drug is metabolized after its administration, and the concentration of a drug is considerably reduced before it reaches to the systemic circulation [34].

### 2.4.4  Parenteral Administration: Intravenous, Subcutaneous, and Intramuscular

In parenteral routes, drugs are inserted straightly across the skin barriers into vascularized tissues (systemic circulation). This kind of administration is preferred especially in emergencies when rapid absorption is necessary or when the patient is not conscious or compliant with taking drug orally. Intravenous delivery is the fastest and the most bioavailable rout of introducing a drug into the body systemic circulation [28]. In this kind of drug administration, the liquefied drug is injected directly into the blood via a sterile injector. In subcutaneous delivery, the liquid

**Table 2.4** Advantages and disadvantages of parenteral routs intravenous, subcutaneous, and intramuscular (Reprinted with modification from [28], Copyright (2012), with permission from Wolters Kluwer Health, Inc.) [28, 29]

| Parenteral Rout | Advantages | Disadvantages |
|---|---|---|
| Intravenous (IV) | Infusion of the largest volumes of drug is possible<br>Administration of a more accurate dose<br>Rapid response is available as soon as vein is accessed<br>Suitable for drugs containing irritating agents<br>Blood sampling through the line is available | Systemic infection<br>Requires trained personnel in a clinical setting (such as an infusion center)<br>Difficulty of needle insertion<br>More costs in comparison with the SC route<br>Slow workflow<br>Requires more observation after injection for diagnosis side effects<br>Possibility of damage surrounding tissues |
| Intramuscular (IM) | Infusion of a larger volume (2–5 mL) in comparison with SC route<br>Appropriate for not readily soluble drugs | Painful<br>Risk of hitting/damaging nerves and veins<br>Risk of infection, local induration, bleeding, and abscess formation<br>Intimidating for "needle-phobic" patients<br>Requires trained personnel<br>High costs in comparison with the SC route |
| Subcutaneous (SC) | Administration by the patient or caregiver<br>Improved compliance<br>Improved patient quality of life<br>Decreased costs<br>Less painful than either IV or IM routes<br>Low risk of systemic infection<br>Possibility of large number of possible injection sites for multiple dosing | Small (1–2 mL) volume of drug<br>Possibility of degradation at injection site (decreased bioavailability)<br>Possibility of retention at injection site (local adverse reactions with irritating compounds) |

drug is administered in subcutaneous tissue, while in intramuscular delivery, the liquid drug is administered in the muscle tissue by injecting with an injector [29, 30]. Table 2.4 indicates the advantages and disadvantages of each parenteral rout.

## 2.5   Classification of DDSs Based on Physical Properties of Dosage Forms

We can classify different types of DDSs according to their physical state of dosage forms. Typically, dosage forms can be categorized into four types including gaseous (such as anesthetics), liquid (like solutions, emulsions, and suspensions), semisolid (such as gels, creams, ointments, and pastes), and solid dosage forms (like powders, granules, and tablets). Most of the dosage forms have several phases (a volume element of a system that is separated from another volume in that system by an interface or phase boundary, due to the immiscibility of the matter in the matter on the other side of the boundary). If the phases in the system are in the same physical state that is an emulsion, formed from two liquid phases (oil and water), although the states of these phases are similar, they differ in their physical properties (density and electrical conductivities), so they tend to be separated from each other by an interface. Suspension is an example of a dosage form that contains phases of different states (a liquid and a solid phase) [1].

From the formulation viewpoint, a dispersed phase leads to physical instability in the system which is undesirable. This problem can be solved by adding emulsifiers to the systems (emulsion is pharmaceutically stabilized) [1].

## 2.6   Classification of DDSs Based on Drug-Loading Mechanism

Drug loading is defined by the encapsulation efficiency, which is described as the fraction of the drug added to the carrier [31]. There are a few reasons why drugs may need to be encapsulated—in general, this is done to reduce the effect of their high toxicity, improve solubility, protect the drug from external conditions (pH, enzymes, and oxidation), and finally to improve circulation time and availability. The encapsulating vehicle can at the same time act as a targeting platform, to help deliver the drug to specific cells in the body [32].

To obtain a precise dose drug inside the delivery device, two categories of loading methods can be employed: (1) direct processing, also known as in situ drug loading, and (2) indirect processing, so-called ex situ drug loading. The advantages and disadvantages of each method have been reviewed elsewhere [33]. In direct processing (in situ loading), a drug is incorporated inside the carrier during the carrier producing. Indirect processing (ex situ loading) consists of a two-step preparation: first, fabrication of the carrier and second, impregnation of the carrier by the drug solution/suspension [33].

# References

1. Perrie Y, Rades T (2009) Pharmaceutics: drug delivery and targeting, 1st edn. Pharmaceutical Press
2. Jain N, Valli K, Vk Devi (2010) Importance of novel drug delivery systems in herbal medicines. Pharmacogn Rev 4:27. https://doi.org/10.4103/0973-7847.65322
3. Jain KK (2008) Drug delivery systems—an overview. In: Jain KK (ed) Drug delivery systems. Humana Press, pp 1–50
4. Das D, Pal S (2015) Modified biopolymer-dextrin based crosslinked hydrogels: application in controlled drug delivery. RSC Adv 5:25014–25050. https://doi.org/10.1039/C4RA16103C
5. Webster John G (2006) Encyclopedia of medical devices and instrumentation, 2nd edn. Wiley, Hoboken, NJ, USA
6. Agrawal S, Giri TK, Tripathi DK, Alexander AA (2012) A review on novel therapeutic strategies for the enhancement of solubility for hydrophobic drugs through lipid and surfactant based self micro emulsifying drug delivery system: a novel approach. Am J Drug Discov Dev 2:143–183. https://doi.org/10.3923/ajdd.2012.143.183
7. Anselmo AC, Mitragotri S (2014) An overview of clinical and commercial impact of drug delivery systems. J Control Release 190:15–28. https://doi.org/10.1016/j.jconrel.2014.03.053
8. Jain KK (2014) Current status and future prospects of drug delivery systems. In: Jain KK (ed) Drug delivery system. Humana Press, New York, NY, pp 1–56
9. Langer RS, Peppas NA (1981) Present and future applications of biomaterials in controlled drug delivery systems. Biomaterials 2:201–214. https://doi.org/10.1016/0142-9612(81)90059-4
10. Coelho JF, Ferreira PC, Alves P, Cordeiro R, Fonseca AC, Góis JR, Gil MH (2010) Drug delivery systems: advanced technologies potentially applicable in personalized treatments. EPMA J 1:164–209. https://doi.org/10.1007/s13167-010-0001-x
11. Chien Y e W (1991) Novel drug delivery systems, 2nd edn. CRC Press, Boca Raton, Florida
12. Chien YW, Lin S (2007) Drug delivery: controlled release. Encycl Pharm Technol 1082–1103
13. Chien YW, Lin S (2002) Optimisation of treatment by applying programmable rate-controlled drug delivery technology. Clin Pharmacokinet 41:1267–1299. https://doi.org/10.2165/00003088-200241150-00003
14. Heller J (1993) Modulated release from drug delivery devices. Crit Rev Ther Drug Carrier Syst 10:253–305
15. Hillery AM, Park K (2017) Drug delivery: fundamentals & applications, 2nd edn. CRC Press
16. Keraliya RA, Patel C, Patel P, Keraliya V, Soni TG, Patel RC, Patel MM (2012) Osmotic drug delivery system as a part of modified release dosage form. ISRN Pharm 2012:1–9. https://doi.org/10.5402/2012/528079
17. Verma RK, Mishra B, Garg S (2000) Osmotically controlled oral drug delivery. Drug Dev Ind Pharm 26:695–708. https://doi.org/10.1081/DDC-100101287
18. Djupesland PG (2013) Nasal drug delivery devices: characteristics and performance in a clinical perspective—a review. Drug Deliv Transl Res 3:42–62. https://doi.org/10.1007/s13346-012-0108-9
19. Lelawongs P, Jue-Chen L, Siddiqui O, Chien YW (1989) Transdermal iontophoretic delivery of arginine-vasopressin (I): physicochemical considerations. Int J Pharm 56:13–22. https://doi.org/10.1016/0378-5173(89)90055-0

20. Maderuelo C, Zarzuelo A, Lanao JM (2011) Critical factors in the release of drugs from sustained release hydrophilic matrices. J Control Release 154:2–19. https://doi.org/10.1016/j.jconrel.2011.04.002
21. Sood N, Bhardwaj A, Mehta S, Mehta A (2014) Stimuli-responsive hydrogels in drug delivery and tissue engineering. Drug Deliv 23:1–23. https://doi.org/10.3109/10717544.2014.940091
22. Kumar A, Nautiyal U, Kaur C, Goel V, Piarchand N (2017) Targeted drug delivery system: current and novel approach. 5:448–454
23. Bae YH, Park K (2011) Targeted drug delivery to tumors: myths, reality and possibility. J Control Release 153:198–205. https://doi.org/10.1016/j.jconrel.2011.06.001
24. Desai PP, Date AA, Patravale VB (2012) Overcoming poor oral bioavailability using nanoparticle formulations—opportunities and limitations. Drug Discov Today Technol 9: e87–e95. https://doi.org/10.1016/j.ddtec.2011.12.001
25. Smart JD (2005) Buccal drug delivery. Expert Opin Drug Deliv 2:507–517. https://doi.org/10.1517/17425247.2.3.507
26. Hao J, Heng PWS (2003) Buccal delivery systems. Drug Dev Ind Pharm 29:821–832. https://doi.org/10.1081/DDC-120024178
27. Jannin V, Lemagnen G, Gueroult P, Larrouture D, Tuleu C (2014) Rectal route in the 21st century to treat children. Adv Drug Deliv Rev 73:34–49. https://doi.org/10.1016/j.addr.2014.05.012
28. Dychter SS, Gold DA, Haller MF (2012) Subcutaneous drug delivery. J Infus Nurs 35:154–160. https://doi.org/10.1097/NAN.0b013e31824d2271
29. Doyle GR, McCutcheon AJ (2012) Clinical procedures for safer patient care. British Columbia Institute of Technology (BCIT)
30. Ard GHW (1998) Formulation-related problems associated with intravenous drug delivery. J Pharm Sci 87:787–796. https://doi.org/10.1021/js980051i
31. Wallace SJ, Li J, Nation RL, Boyd BJ (2012) Drug release from nanomedicines: selection of appropriate encapsulation and release methodology. Drug Deliv Transl Res 2:284–292. https://doi.org/10.1007/s13346-012-0064-4
32. Kita K, Dittrich C (2011) Drug delivery vehicles with improved encapsulation efficiency: taking advantage of specific drug-carrier interactions. Expert Opin Drug Deliv 8:329–342. https://doi.org/10.1517/17425247.2011.553216
33. Parent M, Baradari H, Champion E, Damia C, Viana-Trecant M (2017) Design of calcium phosphate ceramics for drug delivery applications in bone diseases: a review of the parameters affecting the loading and release of the therapeutic substance. J Control Release 252:1–17. https://doi.org/10.1016/j.jconrel.2017.02.012
34. Pond SM, Tozer TN (1984) First-pass elimination. Clin Pharmacokinet 9:1–25. https://doi.org/10.2165/00003088-198409010-00001

# Chapter 3
# Drug Modification

## 3.1 Introduction

Physical, chemical, and biological properties of a drug carrier determine how it can obtain to its predetermined aim. In this chapter, some of these features are evaluated. First, surface characters of the carrier are evaluated, and then the effect of size, shape, and elasticity is explained.

## 3.2 Surface Character

Each particle, depending on the depth that it must penetrate to do its therapeutic function, faces with different kinds of barriers consisting of harsh environment of blood, immunological systems, structural barriers to enter extracellular space, endocytosis process, escaping from lysosome, harsh environment of cytosol, and crossing membrane of cell core [1]. Overall, these barriers can be divided into four general categories: (1) filtering organs such as lung and liver, (2) the physical dams such as vascular wall and extracellular matrix, (3) cellular including escaping from phagocytosis, and (4) enzymatic that must survive from serum and enzymatic activation in endocytosis process [2].

Most of the drugs possess a hydrophobic character and some of them, especially anticancer drugs, are toxic. To decrease their undesirable side effect and avoiding the aforementioned barriers, surface modification of the drugs or their carrier by ligands and hydrophilic agents has been considered by scientist [3]. In the following, some of these agents and ligands are introduced.

Polyethylene glycol usually is used to modify the drug surface because it increases hydrophilicity of the drug and enables the drug to dissolve in water solutions. In addition, polyethylene glycol possesses low toxicity and flexible polymer chain that enable covalently attaching active targeting moieties to the drug surface [3].

© The Author(s) 2018
H. Reza Rezaie et al., *A Review of Biomaterials and Their Applications in Drug Delivery*, SpringerBriefs in Applied Sciences and Technology,
https://doi.org/10.1007/978-981-10-0503-9_3

Transferrin receptor is one of the biological agents easily available on cancer cells. Transferrin, a glycoprotein, normally attaches to the surface of iron ion and delivers it to specific cells. The ligand has been used as a targeting agent on drug carries [3].

Integrin is a transmembrane receptor that is available on the cell surface and is involved in cell adhesion and many intracellular pathways. One of the ligands that is used to target the cell by this receptor is RGD peptide (Arg-Gly-Asp) [4].

Lectin is another cell receptor that can directly or by water-mediation contact to the hydroxyl domain of carbohydrates. Some of the lectin families are asialogly-coprotein receptor (ASGPR) and mannose [4]. Hence, for instance, sugar complexes that can attach to lectins of the cell surface also can be used as a targeting agent [3].

Vitamins are the most important organic compounds that facilitate body functions. Because of the high rate of proliferation in cancerous cells, the high amount of vitamins is essential and these cells overexpress the required receptors to uptake vitamins. Hence, decorating drug carriers with different vitamins such as vitamin B12, folic acid (a low molecular weight pterin-based vitamins [3]), and biotin is a common method to target cancer cells [4].

Phagocytosis, an immunological response of the body, happens in two steps: at the first step called the opsonization, some plasma proteins adsorb on the surface of an invader, and at the next step, macrophages recognize the decorated particles and uptake them [5]. Antibodies or immunoglobulins, Ig, are common proteins in the body that by binding with antigen characterize the foreign body and let clearance of them by the immunosystem of the body. They have five classes including IgA, IgD, IgE, IgG, and IgM. Some cells overexpress some antigens that can be used as a cue in target drug delivery by detecting them via antibodies decorated carriers [4].

## 3.3  Size, Shape, and Elasticity

Size, shape, and elasticity of a carrier also must be considered in designing to improve its half-life circulation.

For example, blood vessels in the healthy circulatory system are covered by endothelial cells, only let too small things, lower than 7 nm, penetrate through them. So, for most drug carriers that are larger than this size and administrated intravascularly, the carriers must attach to the nearest site to the targeted tissue in the circulatory system and release their compartment to diffusion through endothelial cells and receive the targeted tissue. On the other hand, the high rate of angiogenesis that happens in some diseases, especially in cancer, leads to some leakage in the endothelial barrier that lets larger particles to pass it [6].

In addition to penetrate through endothelial cells, the carrier adhesion to a cell is influenced by particle size. By increasing particle size, greater than 500 nm, flow of body fluids imposes the ability of particle adhesion to cells, i.e., decreasing adhesion by increasing size [6].

Overall, in the body, particle uptaking depends on its size and nature that can be done by five different routes including clathrin-coated mediated endocytosis (for particles lower than 200 nm), caveolae (for particles with the size of about 50 nm), cholesterol-mediated endocytosis (for particle size in the range of 50–500 nm), phagocytosis (for particle greater than 750 nm), and micropinocytosis (for particle size in the range of 500 nm–5 μm) [2, 6].

The shape of carriers affects their circulation, the spherical one due to uniform distribution of flow pressure on the carrier flies through blood fluid but nonspherical one rotates in blood and its lateral drift influences the margination and adhesion probabilities. In addition, nonspherical particles due to difficulty to adhesion and being uptaken by phagocytes increase life circulation in blood [6].

There is a controversial discussion about the effect of elasticity on internalizing a carrier to a cell; in some cases, harder particles enter more and in some cases softer ones. It causes different properties of different cells and also different mechanisms of uptaking, the harder uptake by clathrin- and caveolae-mediated endocytosis mechanisms and the softer by micropinocytosis [7].

The elasticity of a carrier can affect the amount of drug uptaken by phago-cytes, the harder is uptaken more than the softer. Maybe its elasticity makes a change in the size of the carrier which cannot be detected by phagocytes and skip this barrier [7].

The elasticity can be modulated. For example, the elasticity of a layer-by-layer structure encapsulated drugs and protein can be tuned by the number of layers, cross-linking density, and removing its core. Or for hydrogel formed from hydro-philic polymer chains cross-linked to each other to form a network and encapsulate water, the cross-linking density, and the amount of added chains determine the elasticity of it [7].

# References

1. Veiseh O, Gunn JW, Zhang M (2010) Design and fabrication of magnetic nanoparticles for targeted drug delivery and imaging. Adv Drug Deliv Rev 62:284–304. https://doi.org/10.1016/j.addr.2009.11.002
2. Parodi A, Corbo C, Cevenini A, Molinaro R, Palomba R, Pandolfi L, Agostini M, Salvatore F, Tasciotti E (2015) Enabling cytoplasmic delivery and organelle targeting by surface modification of nanocarriers. Nanomedicine 10:1923–1940. https://doi.org/10.2217/nnm.15.39
3. Yu W, Zhang N (2009) Surface modification of nanocarriers for cancer therapy. Curr Nanosci 5:123–134. https://doi.org/10.2174/157341309788185370
4. Y X, Y D (2009) Recent advances in biological strategies for targeted drug delivery. Cardiovasc Hematol Disord-Drug Targets 9:206–221. https://doi.org/10.2174/187152909789007025
5. Mornet S, Vasseur S, Grasset F, Duguet E (2004) Magnetic nanoparticle design for medical diagnosis and therapy. J Mater Chem 14:2161. https://doi.org/10.1039/b402025a

6. Sen Gupta A (2016) Role of particle size, shape, and stiffness in design of intravascular drug delivery systems: insights from computations, experiments, and nature. Wiley Interdiscip Rev Nanomedicine Nanobiotechnology 8:255–270. https://doi.org/10.1002/wnan.1362
7. Anselmo AC, Mitragotri S (2017) Impact of particle elasticity on particle-based drug delivery systems. Adv Drug Deliv Rev 108:51–67. https://doi.org/10.1016/j.addr.2016.01.007

# Chapter 4
# Controlled Drug Delivery Systems

## 4.1 Introduction

Drug concentration must be placed between two thresholds, referred as the minimum effective concentration and the maximum toxic concentration, to be effective and nontoxic. Usually, a drug is released in three manners (Fig. 4.1): Burst release, at which a drug eventually dissolves in the surrounding medium and releases all of its amount; periodical release, at which a drug is released periodically and preserves its concentration with high fluctuation between the two thresholds; and zero-order release, at which a drug is released in a constant concentration without significant variation for a long time. The zero-order release is the idealized one, but a drug is usually released in three phases: Phase I, burst release due to releasing of the drugs placed close to the surface and water hydrated layer; Phase II, slow-rate release due to controlled releasing rate is happened by the diffusion rate of water in the structure; and Phase III, fast release due to start of the structure degradation. Hence, the ability to control the drug release is crucial in designing a drug carrier [1].

## 4.2 Drug Release Mechanisms

A drug carrier releases its compartment based on three overall mechanisms including water-filled pore, drug diffusion, and erosion:

- Drug release controlled by water-filled pore: In this manner, water diffuses in the structure of a carrier, for example, a polymeric carrier, and makes it swell up. The encapsulated drug can find its way and diffuses outward through enlarged pores due to the concentration gradient from the inner to the outer of the drug carrier [1].
- Drug release controlled by diffusion: In this manner, a drug diffuses through its carrier structure and its rate depends on permeability of the shell membrane [1].

© The Author(s) 2018
H. Reza Rezaie et al., *A Review of Biomaterials and Their Applications in Drug Delivery*, SpringerBriefs in Applied Sciences and Technology,
https://doi.org/10.1007/978-981-10-0503-9_4

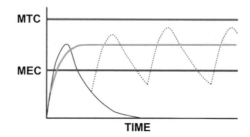

**Fig. 4.1** Three kinds of release kinetic of a drug from a carrier, burst release (black solid line), periodical release (black dashed line), and zero-order release (orange line). The red and blue lines refer to the maximum toxic concentration (MTC) and minimum effective concentration (MEC), respectively (Reprinted with permission from [1]. Copyright (2016) American Chemical Society)

- Erosion: A drug carrier can be eroded in two modes, surface and bulk erosion. In the former, the carrier degrades from the surface to inward faster than water diffusion and releases gradually its drug compartment. In the bulk erosion, water diffusion is faster and a carrier structure degrades from the interior. In this method, the drug is released in less controlled rate than surface erosion [1].

For instance, bulk erosion was observed in the nanoparticles of poly (D,L-lactide-co-glycolide) (PLGA) produced by Jain et al. During 4 weeks, the size of particles started to increase and became 3 times higher than the initial one. Then, the particles fragmented eventually. In these weeks, water gradually diffused in the structure of the polymer, hydrated, and dissolved it, so the particles first swelled and then the structure was destroyed [2].

In addition to the aforementioned overall mechanism, a drug can be released based on two stimuli-responsive classes containing endogenous stimuli, which releases a drug depended on the environmental condition such as pH, enzymes, and so on., and exogenous stimuli, which releases a drug controlled by temperature, light, and ultrasonic stimuli [1, 3].

## 4.2.1  pH-Responsive

pH varies in different points of body, for example, in the stomach is 2, in the intestine is 5–8, near cancer cells change from 5.4 to 7.2 and within endosomal compartment during endocytosis process drops from 6.2 to 5. In different pHs, acids and bases express in suitable ionization state, attach to a carrier, and change its conformation [3].

Zhou et al. synthesized a core–shell structure containing phenolphthalein as a model drug, the core consisting of magnetic nanoparticle $Fe_3O_4$ coated by pH-sensitive poly((2-dimethylamino) ethyl methacrylate) (PDMAEMA). The inorganic coat conformation expanded at pH of 3 because the polymer structure was protonated and its chains electrostatically repulsed each other. On the other hand, the structure compacted again at pH of 7. During 48 h, the drug release from the core–shell structure increased about 30% when pH changed from 7 to 3 [4].

An injectable pentablock copolymer, poly(2-diethylaminoethyl-methyl methacrylate)-poly(ethylene oxide)-poly(propylene oxide)-poly(ethylene oxide)-poly(2-diethylaminoethyl-methyl methacrylate) ($PDEAEM_{25}$-$PEO_{100}$-$PPO_{65}$-$PEO_{100}$-$PDEAEM_{25}$), was synthesized by Determan et al. that could be gelled when the temperature changed from room to physiological temperature. A model drug, nile blue chloride (NBC), was added to the structure. At pH greater than 7.5, hydrophobicity of the structure was dominant but by decreasing pH hydrophilic character took over and water could diffuse to the structure, dissolve it, and release the drug [5].

Tang and Pan synthesized a di-hydrophilic block copolymer, poly(ethylene oxide)-block-poly(glycerol monoacrylate) (PEG-b-PGA), with a model drug, 1-pyrenecarboxaldehyde (Py-CHO), covalently attached to the hydrophobic core of the structure by a pH-sensitive acetal linkage. Release of the drug was evaluated at pH 7.4, physiological pH, and pH 5, endosomal pH. After 72 h, about 80% of the drug was released at pH 5 due to its linkage cleavage; however, at pH 7.4 lower than 10% was released [6].

A polyurethane (PU) microcapsule consisting of hydrophobic polycaprolactone (PCL) and hydrophilic polyethylene glycol (PEG) was produced by Niu et al. to enter the cell and release a drug. The microcapsule compartment contained an anticancer drug, doxorubicin hydrochloride (DOX), and sodium bicarbonate ($NaHCO_3$). Mammalian cells require uptake $HCO_3^-$ from extracellular environment. As shown in Fig. 4.2, during uptaking the microcapsule and $HCO_3^-$, $NaHCO_3$ was protonated and rapidly produced $CO_2$ gas. The produced gas created pores in the surface of the microcapsule and favored burst release of DOX [7].

## 4.2.2 Thermo-Responsive

Thermo-responsive drug release can be classified into both classes of endo- and exogenous stimuli. In endogenous mode, the temperature can differ by abnormal blood flow, high rate of metabolic activation, and high rate of cell proliferation. In the other mode, temperature variation is applied by external sources, such as in hyperthermia treatment [1]. Temperature causes volume transition by changing the intramolecular forces such as Van-der-Waals interaction, hydrophobic interaction, hydrogen bonding, and ionic interaction [3].

Su et al. produced a sandwich nanosheet with a core of $MoS_2$, a photothermal-sensitive agent, coated by mesoporous organosilica. DOX, as a drug model, was loaded into the structure and its release trend was evaluated with/without laser irradiation on the solution containing the nanostructure. Laser irradiation increased the temperature of the solution up to 60°C and enhanced the drug release from lower than 1% to about 7% [8].

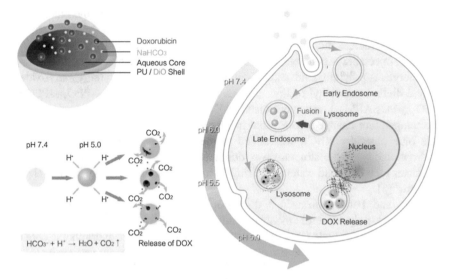

**Fig. 4.2** A schematic of a microcapsule containing doxorubicin and NaHCO₃. The latter in lysosome is protonated, produces gas, and creates pore on the surface of the microcapsule particle to release compartmented drug (Reprinted from [7], Copyright (2017), with permission from Elsevier)

A metal-organic framework consisting of Zr-base, as a metal ion, and (2E,2'E)-3,3'-(naphthalene- 1,4-diyl)diacrylic acid (H₂NPDA), as an organic ligand, was synthesized by Jiang et al. A model drug, diclofenac sodium, was attached to the framework with the naphthalene moiety. After 25 h, the release of the drug by increasing temperature increased from lower than 5% to more than 90% [9].

Dabbagh et al. produced a core–shell nanostructure of mesoporous silica, as a core, and polyacrylamide (PAA), as a protective shell. The gelation temperature of PAA can be manipulated by changing crosslinker/monomer ratio to achieve a thermosensitive shell. DOX was loaded to the structure as a model drug. After heating the nanostructure at 60 °C, the protective shell was melted and revealed the porous core structure. At body temperature, the drug release was about 5% which was enhanced to more than 70% by increasing the temperature to 60 °C [10]. In other research, Debbagh et al. produced another core–shell structure consisting of low melting polyethylene glycol (PEG) with 1500 Da molecular weight. Release of DOX increased from 20 to 70% by increasing the temperature from 37 to 50 °C [11].

Spohr et al. synthesized a membrane from poly(ethylene terephthalate) (PET) foil, and a layer of poly(N-isopropylacrylamide) (NIPAAm) hydrogel was grafted on it. The hydrogel has the ability to change its swollen properties based on temperature variation. The permeability test showed that by increasing the temperature from 25 to 40 °C, permeability of the structure improved about 3 times. By decreasing temperature, the hydrogel swelled and the porous structure was closed [12].

Tetrandrine is a potential anticancer drug that can cover and block the potassium channel of cancer cell and inhibit its proliferation. Shi et al. produced polylactic-co-glycolic acid (PLGA) microparticles encapsulated tetrandrine and $Fe_3O_4$, as heat generator due to its magnetic property. Glass transition temperature ($T_g$) of PLGA is about 44–48 °C. When heating the particles higher than $T_g$, the polymer chains were more mobile and flexible that enhanced the diffusion of the drug molecule through the polymer matrix. The release of tetrandrine increased from 30 to 60% by increasing the temperature from 25 to 45 °C [13].

### 4.2.3   Reactive Oxygen Species

Oxidation response based on reactive oxygen species (ROS), which locally high level of it happens during some diseases such as cancer, can oxidize a carrier and release its compartment [1].

Xiao et al. synthesized a core–shell copolymer of poly(ethylene glycol) diacrylate (PEGDA) and 1,2-ethanedithiol (EDT) by thiol–ene polymerization. The copolymer contained a hydrophilic–hydrophobic property that made it thermal-sensitive. By increasing temperature, the produced nanoparticles collapsed and a drug was encapsulated in the hydrophobic core. At high temperature, the other group easily oxidized and decomposed to hydrophilic parts soluble in water and can release the drug compartment (Fig. 4.3) [14].

Hyper-activation of enzymes during some diseases such as diabetes and cancer leads to creation of ROS that can be used as a stimulus for suitable drugs. Gupta et al. produced a micellar structure consisting of propylene sulfide (PPS) and N,N-dimethylacrylamide (poly($PS_{74}$-b-$DMA_{310}$)). Because PPS is very potent to be oxidized, in an oxidative environment, PPS oxidizes to poly(propylene sulphoxide) and poly(propylene sulphone). Hence, the hydrophobic structure of the micelle decomposed to hydrophilic moiety, and encapsulated hydrophobic compartmented drug was released into the environment [15].

Mesoporous titanium dioxide nanoparticles were produced by Shi et al. and docetaxel was loaded into its porous structure. The mesoporous structure had the

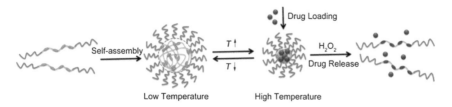

**Fig. 4.3** Schematic of poly(ethylene glycol) diacrylate (PEGDA) core–shell structure that encapsulates a drug by increasing temperature and can release it at high temperature by reaction with reactive oxygen species (Reprinted from [14] with permission of The Royal Society of Chemistry)

ability to produce ROS under ultrasound absorption. So, the author used the feature for a controlled release of the drug by adding a gatekeeper, referred as cyclodextrin (CD). CD is attached to the surface of the nanoparticles by an ROS-sensitive linker and closed the porous structure. By applying ultrasound, the gatekeeper was detached from the surface and compartmented drug was released to the environment [16].

### 4.2.4  Enzyme

Biological processes are controlled by the amount and kinds of enzymes expressed in the surrounding media. A drug carrier can use this feature, especially in targeted drug delivery, to find its way to a predetermined site and release its compartment based on the catalytic reaction of enzyme on the carrier structure [1].

Medina et al. produced poly(amidoamine) dendrimers to carry DOX attached covalently to the substrate by aromatic azo linkers. The drug could be released due to linker cleavage by reaction of azoreductase enzymes of hepatic cytoplasm [17].

Accumulation of degraded products of degradable biomaterials can create inflammation responses. To avoid this problem, an anti-inflammatory drug was added to an electrospun scaffold by Pan et al. Ibuprofen (IBU), as the drug was attached via ester linker to the poly(hydroxyethyl methacrylate) (PHEMA) scaffold. The linker was cleaved by an important regulator of lipid metabolism in vivo, referred as lipase, and controlled inflammatory responses [18].

Chu et al. evaluated drug release from poly(ethylene carbonate) (PEC) in the presence of cholesterol esterase (CE) enzyme. The polymer despite from PLGA did not make the environment acidic during degradation and could maintain its compartment properties. Surface erosion of PEC enables gradual release of the compartment drug, and its kinetic could be improved by the presence of CE enzyme [19].

### 4.2.5  Ultrasound Response

Ultrasound has been used as an enhanced method to deliver drugs, and the method is called sonophoresis. It usually applies in two modes: high-frequency sonophoresis (HFS) (0.7–16 MHz) and low-frequency sonophoresis (LFS) (20–100 kHz). In this method, a coupling medium is applied between an ultrasound probe and skin, whereas its properties such as viscosity, density, acoustic impedance, etc. are the key roles in efficiency of skin permeability. The medium can be included a drug [20].

During sonication, the movement of sound waves displaces particles and changes the local density based on the location places in a low-pressure or a high-pressure cycle of the wave [20]. Transient cavitation is the primary mechanism that happens in sonication. Based on this cavitation, bubbles are created and imploded periodically and after imploding a high-pressure fluid movement, referred as the jet, it enhances penetration through stratum coreneum [21].

Ultrasound has been used in two modes as pretreatment, in which a drug is applied after sonication, and simultaneous treatment, in which a drug is applied during sonication. The second mode is common in HFS, and both methods can be applied in LFS. The selection of the modes depends on the drug, if its structure can be stable during sonication, patient complaint, and penetration efficiency of LFS/HFS [20].

To minimize side effects of a drug used in cancer therapy and to improve its therapeutic application, sustained release of the drug is very important. To this aim, Miyazaki et al. produced an ethylene vinyl alcohol copolymer containing 5-fluorouracil (5-FU) drug and showed by exposing the structure to ultrasound waves, the release of the drug increased more than 10 times compared to no-irradiation condition [22].

Li et al. produced a microbubble, a high molecular weight gas encapsulated into a shell insoluble in water, and added 10-Hydroxycamptothecin (HCPT), a powerful antitumor drug. By exposing the microbubble to ultrasonic waves, accumulation of the drug in cancerous tissue increased 5 times than without ultrasonic waves [23].

Chen and Wu produced a liposome, with diameter of lower than 300 nm, containing a fluorescent agent to evaluate drug release from the liposome under ultrasonic exposure. The fluorescent agent was released about 4 times greater underexposing at 1.1 MHz for 10 s. The increase in agent release could be related to the structural disrupter of the particles greater than 100 nm or pore creation in the structure of particles smaller than 100 nm [24].

## *4.2.6 Magnetic Response*

Scientists have been used magnetic properties of particles in four applications including (1) heating a special site by applying a high-frequency magnetic field on particles, (2) targeting particle conducting with a magnetic field gradient, (3) tissue engineering, and (4) imaging via MRI.

In the targeted drug delivery application, a magnetic field gradient is applied to carry the therapeutic agent toward a predetermined place and the agent is released based on heating the carrier or other aforementioned mechanisms [25].

Among different types of magnetic materials, paramagnetic, diamagnetic, ferromagnetic, and ferrimagnetic materials, the last two kinds are not suitable for targeting drug delivery because magnetization remains even by removing the magnetic field and causes aggregation of particles [25].

Rovers et al. synthesized a polymeric implant with the core of superparamagnetic iron oxide nanoparticle and poly(methyl methacrylate) (PMMA) and the shell of poly(butyl methacrylate-stat-methyl methacrylate) containing ibuprofen. Ibuprofen, besides its therapeutic application, increased $T_g$ of the shell above body temperature. Iron oxide was used as heating agent via magnetic field. By increasing temperature, the shell was disrupted and the drug was released. Due to reversibility of thermo-responsive feature of the implant, by alternatively applying magnetic field, the drug could be released periodically [26].

Hayashi et al. produced a drug carrier including folic acid (FA)- β-cyclodextrin (CD) functionalized superparamagnetic iron oxide nanoparticles. FA was used to bind to a special receptor on breast cancer tumors, referred as folate receptor, and CD was incorporated for two reasons. First, it was a drug container with inner hydrophobic character and outer hydrophilic character that could carry a hydrophobic drug. Second, it is hydrophobic feature of CD depressed by heating. By applying magnetic field, the properties of CD varied and the compartmented drug was released [27].

Fang et al. synthesized poly(lactic-co-glycolic acid) (PLGA) microspheres encapsulating DOX, and iron oxide nanoparticles were attached to the microsphere surface. PLGA at body temperature gradually degraded and caused the drug release but by increasing temperature via exposing magnetic field, the drug release from the microsphere increased 7 times [28].

### 4.2.7  Light Response

In light-responsive drug release, chemical or physical properties of a carrier, such as bonding and conformation, would be imposed under a light source. Many sources of light from high-energy to low-energy have been used but the usage of the high-energy ones, including X-rays and γ-rays, due to catastrophic effects on healthy tissue is abandoned [1].

Niikura et al. produced a nanovesicle containing a shell of gold nanoparticles coated by semifluorinated surface ligands (SFL). The vesicle was dispersible in organic solutions. To use the vesicle in water environment, thiol-terminated poly (ethylene glycol) (dithiol-PEG) was cross-linked to it. By increasing temperature to 62.5 °C, the gap among nanoparticles was opened, let the drug to diffuse into the core of the vesicle and the vesicle encapsulated the drug by decreasing the temperature to room temperature. The structure possessed two features: light-sensitive and the ability to enter the cell. In cell environment and by irradiation of short-term laser (5 min, 532 nm), the structure was opened and released the compartmented drug (the schematic of the carrier and how it encapsulates and releases its compartment are shown in Fig. 4.4) [29].

A nanoscale-coordinated polymer containing hafnium ions and oxygen-responsive singlet linkers, bis-(alkylthio) alkene (BATA), was produced by Liu et al. To stabilize colloidal properties of the nanoparticle, it was covered by PEG. In addition, chlorin e6 (Ce6), a photosensitizer agent, and DOX, a chemotherapeutic drug, were loaded into the structure. The nanostructure under red light exposure (660 nm) produced singlet oxygen, cleaved the BATA linker and caused the drug release [30].

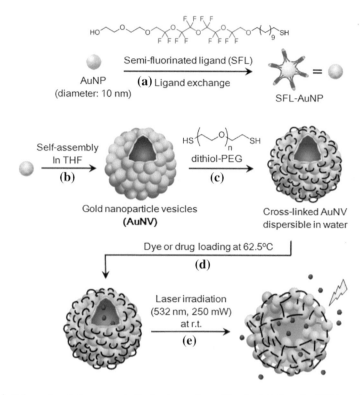

**Fig. 4.4** Schematic of the core–shell structure of semifluorinated ligands (SFL)-coated gold nanoparticle. Thiol-terminated poly(ethylene glycol) (dithiol-PEG) was cross-linked on the shell structure that would open the structure by increasing temperature and exposing to light to encapsulate and release a drug, respectively (Reprinted with permission from [29]. Copyright (2013) American Chemical Society)

Luo et al. produced a stealth liposome containing porphyrin–phospholipid (PoP), a lipid-like molecule, and DOX, a model drug. The structure benefited from two outstanding features: (1) it could improve circulation half-time of the drug to 21.9 h and (2) it could respond to near-infrared light (NIR) and burst release the drug to the cancerous tissue [31].

# References

1. Kamaly N, Yameen B, Wu J, Farokhzad OC (2016) Degradable controlled-release polymers and polymeric nanoparticles: mechanisms of controlling drug release. Chem Rev 116:2602–2663. https://doi.org/10.1021/acs.chemrev.5b00346
2. Jain GK, Pathan SA, Akhter S, Ahmad N, Jain N, Talegaonkar S, Khar RK, Ahmad FJ (2010) Mechanistic study of hydrolytic erosion and drug release behaviour of PLGA nanoparticles: influence of chitosan. Polym Degrad Stab 95:2360–2366. https://doi.org/10.1016/j.polymdegradstab.2010.08.015

3. Schmaljohann D (2006) Thermo- and pH-responsive polymers in drug delivery☆. Adv Drug Deliv Rev 58:1655–1670. https://doi.org/10.1016/j.addr.2006.09.020
4. Zhou L, Yuan J, Yuan W, Sui X, Wu S, Li Z, Shen D (2009) Synthesis, characterization, and controllable drug release of pH-sensitive hybrid magnetic nanoparticles. J Magn Magn Mater 321:2799–2804. https://doi.org/10.1016/j.jmmm.2009.04.020
5. Determan MD, Cox JP, Mallapragada SK (2007) Drug release from pH-responsive thermogelling pentablock copolymers. J Biomed Mater Res Part A 81A:326–333. https://doi.org/10.1002/jbm.a.30991
6. Tang X, Pan C (2008) Double hydrophilic block copolymers PEO- b -PGA: synthesis, application as potential drug carrier and drug release via pH-sensitive linkage. J Biomed Mater Res Part A 86A:428–438. https://doi.org/10.1002/jbm.a.31515
7. Niu Y, Stadler FJ, Song J, Chen S, Chen S (2017) Facile fabrication of polyurethane microcapsules carriers for tracing cellular internalization and intracellular pH-triggered drug release. Colloids Surf B Biointerfaces 153:160–167. https://doi.org/10.1016/j.colsurfb.2017.02.018
8. Su X, Wang J, Zhang J, Yuwen L, Zhang Q, Dang M, Tao J, Ma X, Wang S, Teng Z (2017) Synthesis of sandwich-like molybdenum sulfide/mesoporous organosilica nanosheets for photo-thermal conversion and stimuli-responsive drug release. J Colloid Interface Sci 496:261–266. https://doi.org/10.1016/j.jcis.2017.01.068
9. Jiang K, Zhang L, Hu Q, Zhang Q, Lin W, Cui Y, Yang Y, Qian G (2017) Thermal stimuli-triggered drug release from a biocompatible porous metal-organic framework. Chem A Eur J 23:10215–10221. https://doi.org/10.1002/chem.201701904
10. Dabbagh A, Abdullah BJJ, Abu Kasim NH, Abdullah H, Hamdi M (2015) A new mechanism of thermal sensitivity for rapid drug release and low systemic toxicity in hyperthermia and thermal ablation temperature ranges. Int J Hyperth 31:375–385. https://doi.org/10.3109/02656736.2015.1006268
11. Dabbagh A, Mahmoodian R, Abdullah BJJ, Abdullah H, Hamdi M, Abu Kasim NH (2015) Low-melting-point polymeric nanoshells for thermal-triggered drug release under hyperthermia condition. Int J Hyperth 31:920–929. https://doi.org/10.3109/02656736.2015.1094147
12. Spohr R, Reber N, Wolf A, Alder GM, Ang V, Bashford CL, Pasternak CA, Omichi Hideki, Yoshida M (1998) Thermal control of drug release by a responsive ion track membrane observed by radio tracer flow dialysis. J Control Release 50:1–11. https://doi.org/10.1016/S0168-3659(97)00076-X
13. Shi C, Thum C, Zhang Q, Tu W, Pelaz B, Parak WJ, Zhang Y, Schneider M (2016) Inhibition of the cancer-associated TASK 3 channels by magnetically induced thermal release of Tetrandrine from a polymeric drug carrier. J Control Release 237:50–60. https://doi.org/10.1016/j.jconrel.2016.06.044
14. Xiao C, Ding J, Ma L, Yang C, Zhuang X, Chen X (2015) Synthesis of thermal and oxidation dual responsive polymers for reactive oxygen species (ROS)-triggered drug release. Polym Chem 6:738–747. https://doi.org/10.1039/C4PY01156B
15. Gupta MK, Meyer TA, Nelson CE, Duvall CL (2012) Poly(PS-b-DMA) micelles for reactive oxygen species triggered drug release. J Control Release 162:591–598. https://doi.org/10.1016/j.jconrel.2012.07.042
16. Shi J, Chen Z, Wang B, Wang L, Lu T, Zhang Z (2015) Reactive oxygen species-manipulated drug release from a smart envelope-type mesoporous titanium nanovehicle for tumor sonodynamic-chemotherapy. ACS Appl Mater Interfaces 7:28554–28565. https://doi.org/10.1021/acsami.5b09937
17. Medina SH, Chevliakov MV, Tiruchinapally G, Durmaz YY, Kuruvilla SP, ElSayed MEH (2013) Enzyme-activated nanoconjugates for tunable release of doxorubicin in hepatic cancer cells. Biomaterials 34:4655–4666. https://doi.org/10.1016/j.biomaterials.2013.02.070
18. Pan G, Liu S, Zhao X, Zhao J, Fan C, Cui W (2015) Full-course inhibition of biodegradation-induced inflammation in fibrous scaffold by loading enzyme-sensitive prodrug. Biomaterials 53:202–210. https://doi.org/10.1016/j.biomaterials.2015.02.078

19. Chu D, Curdy C, Riebesehl B, Zhang Y, Beck-Broichsitter M, Kissel T (2013) Enzyme-responsive surface erosion of poly(ethylene carbonate) for controlled drug release. Eur J Pharm Biopharm 85:1232–1237. https://doi.org/10.1016/j.ejpb.2013.04.011

20. Polat BE, Hart D, Langer R, Blankschtein D (2011) Ultrasound-mediated transdermal drug delivery: mechanisms, scope, and emerging trends. J Control Release 152:330–348. https://doi.org/10.1016/j.jconrel.2011.01.006

21. Schoellhammer CM, Blankschtein D, Langer R (2014) Skin permeabilization for transdermal drug delivery: recent advances and future prospects. Expert Opin Drug Deliv 11:393–407. https://doi.org/10.1517/17425247.2014.875528

22. Miyazaki S, Hou W-M, Takada M (1985) Controlled drug release by ultrasound irradiation. Chem Pharm Bull 33:428–431

23. Li P, Zheng Y, Ran H, Tan J, Lin Y, Zhang Q, Ren J, Wang Z (2012) Ultrasound triggered drug release from 10-hydroxycamptothecin-loaded phospholipid microbubbles for targeted tumor therapy in mice. J Control Release 162:349–354. https://doi.org/10.1016/j.jconrel.2012.07.009

24. Chen D, Wu J (2010) An in vitro feasibility study of controlled drug release from encapsulated nanometer liposomes using high intensity focused ultrasound. Ultrasonics 50:744–749. https://doi.org/10.1016/j.ultras.2010.02.009

25. Polyak B, Friedman G (2009) Magnetic targeting for site-specific drug delivery: applications and clinical potential. Expert Opin Drug Deliv 6:53–70. https://doi.org/10.1517/17425240802662795

26. Rovers SA, Hoogenboom R, Kemmere MF, Keurentjes JTF (2012) Repetitive on-demand drug release by magnetic heating of iron oxide containing polymeric implants. Soft Matter 8:1623–1627. https://doi.org/10.1039/C2SM06557F

27. Hayashi K, Ono K, Suzuki H, Sawada M, Moriya M, Sakamoto W, Yogo T (2010) High-frequency, magnetic-field-responsive drug release from magnetic nanoparticle/organic hybrid based on hyperthermic effect. ACS Appl Mater Interfaces 2:1903–1911. https://doi.org/10.1021/am100237p

28. Fang K, Song L, Gu Z, Yang F, Zhang Y, Gu N (2015) Magnetic field activated drug release system based on magnetic PLGA microspheres for chemo-thermal therapy. Colloids Surf B Biointerfaces 136:712–720. https://doi.org/10.1016/j.colsurfb.2015.10.014

29. Niikura K, Iyo N, Matsuo Y, Mitomo H, Ijiro K (2013) Sub-100 nm gold nanoparticle vesicles as a drug delivery carrier enabling rapid drug release upon light irradiation. ACS Appl Mater Interfaces 5:3900–3907. https://doi.org/10.1021/am400590m

30. Liu J, Yang G, Zhu W, Dong Z, Yang Y, Chao Y, Liu Z (2017) Light-controlled drug release from singlet-oxygen sensitive nanoscale coordination polymers enabling cancer combination therapy. Biomaterials 146:40–48. https://doi.org/10.1016/j.biomaterials.2017.09.007

31. Luo D, Carter KA, Razi A, Geng J, Shao S, Giraldo D, Sunar U, Ortega J, Lovell JF (2016) Doxorubicin encapsulated in stealth liposomes conferred with light-triggered drug release. Biomaterials 75:193–202. https://doi.org/10.1016/j.biomaterials.2015.10.027

# Chapter 5
# Nanotechnology in Drug Delivery Systems

## 5.1 Introduction

Because of the harsh and hydrophilic environment of the body, some more potent but poor water-soluble drugs molecule become developed [1]. In addition, without benefit, the advantages of nanodrug delivery, some drugs that need to enter the intracellular space to have suitable therapeutic effects are destroyed in extracellular space with enzymatic degradation, cannot be uptaken by cells or are degraded by interaction with cytosol proteins [2]. By improving nanotechnology and introduction of new drug carriers, the mentioned drawbacks have been solved [1].

One of the abilities of nanomaterials is their mimicking behavior of viruses to penetrate a cell membrane by endocytosis process [3]. To this aim, the nanomaterials must have some features:

- Size and shape to optimize its half-life circulation and entrance to a cell [3]: each particle faces with two elimination systems that threaten its circulation time, including renal clearance and phagocytosis. To avoid former system, the particle must be larger than 10 nm, and the ones smaller than 500 nm cannot be uptaken by phagocytic system [4].
- Surface charges: Based on the cell that is targeted and environmental barrier in front of a carrier, surface charge of the carrier highly impresses uptaking of nanoparticles by cells [3].
- Surface ligand arrangement to conduct a carrier to a predetermined site: The surface properties of a nanoparticle control its solubility and particle–macromolecule interaction. To avoid undesirable interaction which can cause binding nanoparticles to macromolecules and aggregation, nanoparticles are coated with neutral ligands [3].

Furthermore, nanoparticles benefit from their high surface area, because they can carry a large amount of drug and decrease the need to apply a high-dose carrier to reach the therapeutic level [5, 6]. High drug-loading nanomedicine can be divided

© The Author(s) 2018
H. Reza Rezaie et al., *A Review of Biomaterials and Their Applications in Drug Delivery*, SpringerBriefs in Applied Sciences and Technology,
https://doi.org/10.1007/978-981-10-0503-9_5

into four classes based on its carrier, including (1) inert carrier so that the drug is loaded on an inorganic carrier such as mesoporous silica nanoparticles and mesoporous carbon nanoparticles, (2) drug forms part of a carrier such as linear/branched polymer–drug conjugates, (3) drug as a carrier such as drug nanocrystals and amphiphilic drug–drug conjugates, and (4) other complex strategies such as noncovalent assembly and multiple assembly [7].

In the following, some of the common nanostructure carriers in five classes are introduced: nanotubes, mesoporous structures, nanorods, lipids, and quantum dots.

## 5.2   Nanotubes

A carbon nanotube (CNT) is a one-dimensional structure that resembles rolling a graphene sheet like a tube. Its diameter and length are about a few nanometers and maybe a few hundred micrometers, respectively. Some advantages of CNT are as follows: (1) $sp^2$-hybridized structure that can attract drug molecules and targeting agents, (2) high surface area for carrying a large amount of drugs, and (3) the ability to enter cells [8].

For example, Mallekpour and Khodadadzadeh added a biodegradable and biocompatible polymer, starch, to a CNT structure. Then, by adding oleic acid, they could load zolpidem, a hydrophobic anti-insomnia drug, to the structure [9].

Chen et al. also produced a CNT carrier and loaded it by a second-generation taxoid that was covalently added to the structure and could be released inside the cell. In addition, biotin, a tumor-targeting molecule, was attached to the surface of CNT to increase uptaken concentration of the carrier in a specific cell [10].

Hu et al. produced silica nanotube by depositing it on $Gd(OH)_3$ nanorod, as a template. The obtained nanotube was degradable and could decompose to $Si(OH)_2$ that was absorbable by body and eliminated by kidney. The degradation of the silica nanotube depends on its thickness and environmental pH. Because it was nondegradable in low pH and easily can be decomposed at pH of 8, it could be used as oral administration method to deliver its compartment. In addition, by increasing its thickness, release of drug could be tuned [11].

Titanium nanotube was produced by electrochemical anodization of Ti in the form of an array of nanotube perpendicular to the base structure. The structure could be used as an implant drug reservoir and its compartment drug with different methods, mentioned in Chap. 4, consisting of magnetic-, thermal-, pH-, ultrasound-responsive, etc., could be released [12].

Zhou et al. produced an array of titanium oxide nanotube and changed their hydrophilic character to super-hydrophobicity by the immersion of array in a methanol solution of hydrolyzed 1 wt% 1H,1H,2H,2H-perfluorooctyl-triethoxysilane and heated at 140 °C. The super-hydrophobic character at first allowed to attach a hydrophobic drug and then, by trapping an air layer on the array, encapsulated the drug in the structure. The drug then can be released by ultrasonication [13].

Hallowsite nanotube is a natural hollow elongated tubular $Al_2Si_2O_5(OH)_4.nH2O$ that is usually produced during hydrothermal alteration of a rock in wet tropical and volcanic regions. The nanotube containing OH group on inner and outer surfaces of the walls that can be used as connection sites for therapeutic agents. In addition, its structure differs from the inner parts to the outer. At inner, the structure contains $SiO_2$ with negative charge and the outer one contains $Al(OH)_3$ with positive charge that let attaching different molecules to the structure [14].

Tang et al. synthesized calcium carbonate nanotube $CaCO_3$ by supported liquid membrane technology and loaded podophyllotoxin (PPT), an anticancer drug. The nanotube was biodegradable in acidic environment and could release the drug during cellar activity [15].

DNA also has been considered as a nanotube carrier. Its outstanding feature is established on the ability to precisely control nanoscale structure and its natural biocompatibility [16].

## 5.3  Mesoporous Structures

Mesoporous particles are nanoparticles with diameter of 50–300 nm and pore diameter of 3–6 nm [17]. Different kinds of them are introduced below:

One of them is mesoporous silica that despite outstanding features of it such as biocompatibility and osteogenicity, producing highly porous structure of silica, containing a honeycomb-like porous structure, with high surface area and narrow distribution of nanoscaled pore sizes have been considered [18].

Tang et al. produced mesoporous bioactive glass via a bi-phase delamination method and templating by hexadecyl trimethyl ammonium bromide (CTAB) [19].

Knezevict and Lin synthesized a core–shell structure consisting of magnetic iron oxide, as core, and a mesoporous shell of silica. An anticancer drug, camptothecin, was loaded into the porous shell, with pore diameter of 2.6 nm, and 2-nitro-5-mercaptobenzyl alcohol functionalized CdS was attached on the particles by a photocleavable carbamate linkage, as a gatekeeper. The result showed that irradiation a low-power UV light (365 nm) could uncover the porous structure and released the encapsulated drug [20].

Wu et al. produced a mesoporous bioactive glass particle and loaded it with bovine serum albumin (BSA). Due to the bioactivity of the particles, the particles were immersed in SBF solution to precipitate hydroxyapatite on them. The thickness of the hydroxyapatite coating was tailored to control the release of BSA [21].

McMaster et al. synthesized a mesoporous titanium dioxide network by adding on collagen as template. They mixed collagen and a liquid metal oxide precursor and created the hybrid structure via sol–gel method. By calcination, collagen was removed and left a network consisting of walls with 300 nm in thickness and pore diameter of 4.2–8.8 nm. The structure was loaded with ibuprofen and its release profile showed stable trend after 96 h [22].

A mesoporous Fe/carbon composite was produced by Yuan et al. In this research, they produced the structure by adding the composite to a silicon template. The final mesoporous structure consists of pores of 3.1–3.8 nm and wall thickness of 5.7–6.2 nm, and because of the presence of iron, it possessed superparamagnetic properties [23].

Guo et al. produced mesoporous particles of carbonated hydroxyapatite and loaded it with an antibiotic agent, gentamicin. The particles were used as a filler of bone defects. The antibiotic agent prevents bacterial adhesion and biofilm formation. Hence, the particles could accomplish their task as osteoinduction and osteoconductive agents [24].

Bakhtiari et al. produced a mesoporous hydroxyapatite by the means of soft template technique. In this method, cetyltrimethylammonium bromide (CTAB) was used as templating surfactant and an expander agent, 1-dodecanethiol (RSH), also was added. Hydrophobic part of CTAB encapsulated RSH in a micellar structure of CTAB and hydroxyapatite formed on the surface of micelle. It was shown that the pore size of the mesoporous structure depended on the pore expander content, temperature of synthesis process, and expander agent-to-surfactant mass ratio [25–27].

## 5.4  Nanorods

Yu et al. produced porous iron oxide nanorods with 40–60 nm length. Doxorubicin (DOX), an anticancer drug, was loaded into the porous structure and the surface of the nanorods was coated by poly(ethylene glycol)-bisamine ($NH_2$–PEG–$NH_2$), which improved targeting and uptaking properties of them. Its uptaking was confirmed by Hela cells and showed a stable drug release trend [28].

Mirza also synthesized gold nanorods as a carrier for DOX. To improve the nanorods uptake, they were coated with folic acid. In addition, the drug was covalently connected to the surface of nanorods by thiotic acid [29].

Zhang et al. produced mesoporous strontium hydroxyapatite nanorods that benefited from two features, both as a drug carrier and as a self-activated luminescent. The nanorods consisted of 120–150 nm length, 20 nm diameter, and 3–5 nm pore diameter. In addition, it illuminated in the range of 360–570 nm. Ibuprofen was added to the structure as a model drug showed stable release during 35 h [30].

Liu et al. synthesized mesoporous gadolinium oxide nanorods and coated it with poly(ethylene glycol) (PEG). They also loaded camptothecin (CPT) drug into the pore structure. The obtained nanorods had two potentials: it created contrast for imaging evaluation and released chemotherapeutic agent [31].

## 5.5 Lipids

Lipid nanoparticles are divided into two groups: solid lipid nanoparticle and nanostructured lipid carriers. The former outstands with some features containing photo-, moisture-, and chemical-sensitive, easy production with low cost, and avoiding the use of organic solvent. But its crystal structure limits its drug-loading capacity; during storage, it forms a perfect crystal lattice that may cause leakage of encapsulated drug and its drug release profile and some other features may change during storage. To overcome the mentioned drawbacks, nanostructures lipid carriers have been introduced. Many different lipid molecules can be used to maintain and increase imperfection of the structure as much as possible. The imperfection enhances the drug-loading capacity [32].

In each blink, a mucus layer covers ocular surface, removes any microorganism, and plays its role as a barrier for hydrophobic drug molecules to reach to their target tissue. Diebold et al. produced a hybrid of liposome–chitosan nanoparticles that benefited from features of both of their parts. Chitosan was used as a carrier for ocular drug because it adhered to ocular mucosa and entered to superficial layer of cornea and conjunctiva. Liposomes also could carry different types of drugs because of its hydrophilic–hydrophobic structure. So, a complex of them could be used in the form of eye drop to administer any drugs [33].

Wang et al. produced liposome nanoparticles containing poly(D,L-lactide-co-glycolide) (PLGA) as core and PEG and folic acid as outer surface coating. PLGA was selected because of its biodegradability and ability to encapsulate high amount of drug. The coating can interact electrostatically by DNA to co-deliver drug and gene to be used as target delivery and increase circulation time [34].

Hou et al. produced mesoporous CuS nanoparticles and applied them in photothermal therapy due to their photosensitizer and in photodynamic therapy because of their ability to create reactive oxygen species under irradiation of NIR. Transferrin was also added to the nanoparticle surface as a cell marker for tumor detection. In addition, artesunate as an anticancer model drug was encapsulated into the nanoparticles [35].

Parveen and Sahoo synthesized PLGA nanoparticles encapsulating paclitaxel, an anticancer drug. To improve half circulation life of the carrier, it was coated with chitosan and PEG. Chitosan could adhere to mucous membrane and enhance crossing the membrane. Steric repulsion and hydrated outer shell were caused by PEG that avoided the nanoparticles to be uptaken by reticuloendothelial system and improved their circulation lifetime [36].

## 5.6   Quantum Dots

Quantum dots or fluorescent semiconductor nanoparticles are extremely small particles in the range of 2–10 nm. Their sizes are approximately around the dimension of large proteins and smaller than conventional organic and inorganic nanoparticles. The fluorescent property of quantum dots enables them as imaging and tracing agents. Quantum dots can be encapsulated in other nanoparticles and improve their bioimaging and therapeutic features [37].

Chen et al. produced a core–shell nanocarrier, with the core of conjugation of ZnO quantum dots and gold nanoparticles, and the shell of amphiphilic hyper-branched block copolymer consisted of poly(l-lactide) (PLA) inner arm and folate (FA)-conjugated sulfated polysaccharide from gynostemma pentaphyllum makino (GPPS-FA) outer arm. ZnO is a nontoxic quantum dot that could be used as bioimaging agent. Gold nanoparticles were applied as light-stimuli agents for photothermal therapy. The shell copolymer benefited from two features: it enhanced cell entering property of the nanocarrier and also was applied as an anticancer agent [38].

Xu et al. synthesized a core–shell nanoparticle consisting of Mn-doped ZnS quantum dot as core and glycopolypeptides as shell. The shell offered rich sites of –OH and –$NH_2$ that could be used as binding site for a drug, in this case ibuprofen. The drug release showed sustained trend during 700 min [39].

Chen et al. produced carbon nanotube carriers where $Fe_3O_4$ was attached to the inside wall of the nanotube, and $SiO_2$-coated CdTe quantum dot was bonded to the external surface. In addition, transferrin was coated on the nanotube. $Fe_3O_4$ improved carrier delivery to the specific site by the means of magnetic targeting. Quantum dot created the ability of optical imaging of a target tissue, and transferrin enhanced uptake of the carrier and delivery DOX into the interior part of a cell [40].

## References

1. Shi J, Votruba AR, Farokhzad OC, Langer R (2010) Nanotechnology in drug delivery and tissue engineering: from discovery to applications. Nano Lett 10:3223–3230. https://doi.org/10.1021/nl102184c
2. Mailänder V, Landfester K (2009) Interaction of nanoparticles with cells. Biomacromol 10:2379–2400. https://doi.org/10.1021/bm900266r
3. Verma A, Stellacci F (2010) Effect of surface properties on nanoparticle cell interactions. Small 6:12–21. https://doi.org/10.1002/smll.200901158
4. Webster DM, Sundaram P, Byrne ME (2013) Injectable nanomaterials for drug delivery: carriers, targeting moieties, and therapeutics. Eur J Pharm Biopharm 84:1–20. https://doi.org/10.1016/j.ejpb.2012.12.009
5. Alkilany AM, Thompson LB, Boulos SP, Sisco PN, Murphy CJ (2012) Gold nanorods: their potential for photothermal therapeutics and drug delivery, tempered by the complexity of their biological interactions. Adv Drug Deliv Rev 64:190–199. https://doi.org/10.1016/j.addr.2011.03.005

6. Rezaie HR, Bakhtiari L, Öchsner A (2015) Biomaterials and their applications. Springer International Publishing, Cham

7. Shen S, Wu Y, Liu Y, Wu D (2017) High drug-loading nanomedicines: progress, current status, and prospects. Int J Nanomedicine 12:4085–4109. https://doi.org/10.2147/IJN.S132780

8. Meng L, Zhang X, Lu Q, Fei Z, Dyson PJ (2012) Single walled carbon nanotubes as drug delivery vehicles: targeting doxorubicin to tumors. Biomaterials 33:1689–1698. https://doi.org/10.1016/j.biomaterials.2011.11.004

9. Mallakpour S, Khodadadzadeh L (2018) Ultrasonic-assisted fabrication of starch/MWCNT-glucose nanocomposites for drug delivery. Ultrason Sonochem 40:402–409. https://doi.org/10.1016/j.ultsonch.2017.07.033

10. Chen J, Chen S, Zhao X, Kuznetsova LV, Wong SS, Ojima I (2008) Functionalized single-walled carbon nanotubes as rationally designed vehicles for tumor-targeted drug delivery. J Am Chem Soc 130:16778–16785. https://doi.org/10.1021/ja805570f

11. Hu K, Hsu K, Yeh C (2010) pH-Dependent biodegradable silica nanotubes derived from Gd (OH)3 nanorods and their potential for oral drug delivery and MR imaging. Biomaterials 31:6843–6848. https://doi.org/10.1016/j.biomaterials.2010.05.046

12. Losic D, Aw MS, Santos A, Gulati K, Bariana M (2015) Titania nanotube arrays for local drug delivery: recent advances and perspectives. Expert Opin Drug Deliv 12:103–127. https://doi.org/10.1517/17425247.2014.945418

13. Zhou J, Frank MA, Yang Y, Boccaccini AR, Virtanen S (2018) A novel local drug delivery system: superhydrophobic titanium oxide nanotube arrays serve as the drug reservoir and ultrasonication functions as the drug release trigger. Mater Sci Eng C 82:277–283. https://doi.org/10.1016/j.msec.2017.08.066

14. Hanif M, Jabbar F, Sharif S, Abbas G, Farooq A, Aziz M (2016) Halloysite nanotubes as a new drug-delivery system: a review. Clay Miner 51:469–477. https://doi.org/10.1180/claymin.2016.051.3.03

15. Tang J, Sun D-M, Qian W, Zhu R, Sun X, Wang W, Li K, Wang S-L (2012) One-step bulk preparation of calcium carbonate nanotubes and its application in anticancer drug delivery. Biol Trace Elem Res 147:408–417. https://doi.org/10.1007/s12011-012-9325-9

16. Liang L, Shen J, Wang Q (2017) Molecular dynamics study on DNA nanotubes as drug delivery vehicle for anticancer drugs. Colloids Surf B Biointerfaces 153:168–173. https://doi.org/10.1016/j.colsurfb.2017.02.021

17. Colilla M, González B, Vallet-Regí M (2013) Mesoporous silicananoparticles for the design of smart delivery nanodevices. Biomater Sci 1:114–134. https://doi.org/10.1039/C2BM00085G

18. Slowing I, Viveroescoto J, Wu C, Lin V (2008) Mesoporous silica nanoparticles as controlled release drug delivery and gene transfection carriers☆. Adv Drug Deliv Rev 60:1278–1288. https://doi.org/10.1016/j.addr.2008.03.012

19. Tang J, Chen X, Dong Y, Fu X, Hu Q (2017) Facile synthesis of mesoporous bioactive glass nanospheres with large mesopore via biphase delamination method. Mater Lett 209:626–629. https://doi.org/10.1016/j.matlet.2017.08.033

20. Knežević NŽ, Lin VS-Y (2013) A magnetic mesoporous silica nanoparticle-based drug delivery system for photosensitive cooperative treatment of cancer with a mesopore-capping agent and mesopore-loaded drug. Nanoscale 5:1544. https://doi.org/10.1039/c2nr33417h

21. Wu C, Zhang Y, Ke X, Xie Y, Zhu H, Crawford R, Xiao Y (2010) Bioactive mesopore-glass microspheres with controllable protein-delivery properties by biomimetic surface modification. J Biomed Mater Res, Part A 95A:476–485. https://doi.org/10.1002/jbm.a.32873

22. McMaster WA, Wang X, Caruso RA (2012) Collagen-templated bioactive titanium dioxide porous networks for drug delivery. ACS Appl Mater Interfaces 4:4717–4725. https://doi.org/10.1021/am301093k

23. Yuan X, Xing W, Zhuo S, Han Z, Wang G, Gao X, Yan Z (2009) Preparation and application of mesoporous Fe/carbon composites as a drug carrier. Microporous Mesoporous Mater 117:678–684. https://doi.org/10.1016/j.micromeso.2008.07.039

24. Guo Y-J, Long T, Chen W, Ning C, Zhu Z, Guo Y (2013) Bactericidal property and biocompatibility of gentamicin-loaded mesoporous carbonated hydroxyapatite microspheres. Mater Sci Eng C 33:3583–3591. https://doi.org/10.1016/j.msec.2013.04.021

25. Bakhtiari L, Rezaie HR, Javadpour J, Erfan M, Shokrgozar MA (2015) The effect of synthesis parameters on the geometry and dimensions of mesoporous hydroxyapatite nanoparticles in the presence of 1-dodecanethiol as a pore expander. Mater Sci Eng C 53:1–6. https://doi.org/10.1016/j.msec.2015.01.083

26. Bakhtiari L, Javadpour J, Rezaie HR, Erfan M, Shokrgozar MA (2015) The effect of swelling agent on the pore characteristics of mesoporous hydroxyapatite nanoparticles. Prog Nat Sci Mater Int 25:185–190. https://doi.org/10.1016/j.pnsc.2015.06.005

27. Bakhtiari L, Javadpour J, Rezaie HR, Erfan M, Mazinani B, Aminian A (2016) Pore size control in the synthesis of hydroxyapatite nanoparticles: the effect of pore expander content and the synthesis temperature. Ceram Int 42:11259–11264. https://doi.org/10.1016/j.ceramint.2016.04.041

28. Yu P, Xia X, Wu M, Cui C, Zhang Y, Liu L, Wu B, Wang C, Zhang L, Zhou X, Zhuo R, Huang S (2014) Folic acid-conjugated iron oxide porous nanorods loaded with doxorubicin for targeted drug delivery. Colloids Surf B Biointerfaces 120:142–151. https://doi.org/10.1016/j.colsurfb.2014.05.018

29. Mirza AZ (2015) A novel drug delivery system of gold nanorods with doxorubicin and study of drug release by single molecule spectroscopy. J Drug Target 23:52–58. https://doi.org/10.3109/1061186X.2014.950667

30. Zhang C, Li C, Huang S, Hou Z, Cheng Z, Yang P, Peng C, Lin J (2010) Self-activated luminescent and mesoporous strontium hydroxyapatite nanorods for drug delivery. Biomaterials 31:3374–3383. https://doi.org/10.1016/j.biomaterials.2010.01.044

31. Liu Z, Liu X, Yuan Q, Dong K, Jiang L, Li Z, Ren J, Qu X (2012) Hybrid mesoporous gadolinium oxide nanorods: a platform for multimodal imaging and enhanced insoluble anticancer drug delivery with low systemic toxicity. J Mater Chem 22:14982. https://doi.org/10.1039/c2jm31100c

32. Das S, Chaudhury A (2011) Recent advances in lipid nanoparticle formulations with solid matrix for oral drug delivery. AAPS PharmSciTech 12:62–76. https://doi.org/10.1208/s12249-010-9563-0

33. Diebold Y, Jarrín M, Sáez V, Carvalho ELS, Orea M, Calonge M, Seijo B, Alonso MJ (2007) Ocular drug delivery by liposome–chitosan nanoparticle complexes (LCS-NP). Biomaterials 28:1553–1564. https://doi.org/10.1016/j.biomaterials.2006.11.028

34. Wang H, Zhao P, Su W, Wang S, Liao Z, Niu R, Chang J (2010) PLGA/polymeric liposome for targeted drug and gene co-delivery. Biomaterials 31:8741–8748. https://doi.org/10.1016/j.biomaterials.2010.07.082

35. Hou L, Shan X, Hao L, Feng Q, Zhang Z (2017) Copper sulfide nanoparticle-based localized drug delivery system as an effective cancer synergistic treatment and theranostic platform. Acta Biomater 54:307–320. https://doi.org/10.1016/j.actbio.2017.03.005

36. Parveen S, Sahoo SK (2011) Long circulating chitosan/PEG blended PLGA nanoparticle for tumor drug delivery. Eur J Pharmacol 670:372–383. https://doi.org/10.1016/j.ejphar.2011.09.023

37. Probst CE, Zrazhevskiy P, Bagalkot V, Gao X (2013) Quantum dots as a platform for nanoparticle drug delivery vehicle design. Adv Drug Deliv Rev 65:703–718. https://doi.org/10.1016/j.addr.2012.09.036

38. Chen T, Zhao T, Wei D, Wei Y, Li Y, Zhang H (2013) Core–shell nanocarriers with ZnO quantum dots-conjugated Au nanoparticle for tumor-targeted drug delivery. Carbohydr Polym 92:1124–1132. https://doi.org/10.1016/j.carbpol.2012.10.022

39. Xu Z, Li B, Tang W, Chen T, Zhang H, Wang Q (2011) Glycopolypeptide-encapsulated Mn-doped ZnS quantum dots for drug delivery: fabrication, characterization, and in vitro assessment. Colloids Surf B Biointerfaces 88:51–57. https://doi.org/10.1016/j.colsurfb.2011.05.055

40. Chen M-L, He Y, Chen X, Wang J (2012) Quantum dots conjugated with $Fe_3O_4$ -filled carbon nanotubes for cancer-targeted imaging and magnetically guided drug delivery. Langmuir 28:16469–16476. https://doi.org/10.1021/la303957y

# Chapter 6
# 3D Printing Technologies for Drug Delivery

## 6.1 Introduction

As it was mentioned in the previous chapters, the efficiency of medication is dependent on several parameters such as drug release profile, drug transport mechanisms, and most importantly the interaction between a drug and its surrounding environment in the human body. So, scientists have surveyed many methods to improve the drug capsules in order to ameliorate the drug release behavior as well as the drug transport mechanism. These tactics are as follows: (1) surface modification of drug particles, (2) attaching specific functional groups to a drug to improve the interaction between the drug and targeted tissue or cells, and (3) coating the drug with special polymers like PEG in order to deceive the immune system of the body that increases biological half-life time of the drug in blood circulation [1].

In addition to the above methods, recently great attention has been focused to control the three-dimensional (3D) structure and designing different 3D shapes in nano- and microscale dimensions of drug particles. It is proved that controlling the shape and structure of drug particles can increase the efficiency of drugs, especially those designed to be administrated orally that go through the GI tract. Scientists suggest three major reasons to prove their claim: (1) the more complexity of 3D shapes, the more specific surface area that enhances absorption of low soluble drugs; (2) each special 3D shape can be employed to target special cells; and (3) each 3D shape has its unique profile of drug release [1].

In this chapter, we briefly discuss how the shape of drug particles and drug carriers can influence and enhance the drug delivery behaviors. Then, the new technologies currently utilized to fabricate controlled nano- and microstructure shapes are reviewed.

© The Author(s) 2018
H. Reza Rezaie et al., *A Review of Biomaterials and Their Applications in Drug Delivery*, SpringerBriefs in Applied Sciences and Technology,
https://doi.org/10.1007/978-981-10-0503-9_6

## 6.2   The Effect of 3D Shape and Structure on Nano- and Micro-carrier Properties

Although there are numerous documents that explain the effect of particle shape on their performance in drug delivery system, final impact of particle shape on the drug release properties is not yet completely clear. Here, we describe some effects of the particle shape on the drug delivery behaviors.

### 6.2.1   Effect of Carrier Shape Designing on Drug Delivery Performance in Targeting Certain Organs

The first effect of the shape of particle is its role in improving the particle perfor-mance in targeting certain organs; for example, according to the contents in the preceding chapters, some types of insulins and nucleic acids cannot be adminis-trated orally. Because, these biologically active substances are very sensitive in the digestive system and, in addition, their absorption rate is very low due to permanent flow of food and water in the digestive system. Therefore, scientists have tried to increase their stability in the digestive system for the longer time by designing and fabricating them in microneedle shapes that enhance the adhesion of these devices to the surface of the GI tract [2] (Fig. 6.1). This microneedle structure has been used in other administration routs such as transdermal DDSs that has increased the permeability of drugs by four times as compared with conventional DDSs. Additionally, using these microneedle structures does not cause or suffer physical pain because their size is so small which does not reach the nervous system. However, their length is enough large that enable them to penetrate through the skin of the body and stimulate antigen-presenting cells. Hence, they can be used for vaccine injection too. Furthermore, these microneedles can also be manufactured

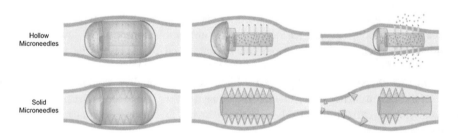

**Fig. 6.1** Schematic graph of microneedle pill for GI targeting. Both hollow and solid microneedles could be used. In both cases, the pill's needles are initially coated by a pH-responsive coating to aid in ingestion. The coating dissolves when the pills reach to its target in GI tract and reveals the microneedles (Reprinted from [2], Copyright (2008), with permission from Elsevier)

**Drug tip-loaded microneedle arrays
fabricated from hyaluronic acid**

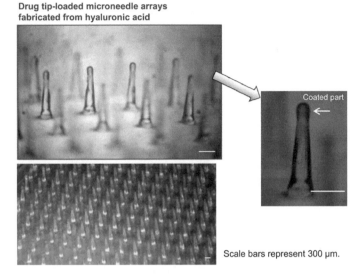

Scale bars represent 300 μm.

**Fig. 6.2** Micrograph of array and single tip-loaded microneedle [3] (Reprinted with permission from [3]. Copyright (2016) American Chemical Society)

from biodegradable polymers loaded with drug, so a sustain drug release is established inside the skin. In Fig. 6.2, a micrograph of a microneedle array loaded by drug on their tip is indicated. In addition to the mentioned advantages, other benefits of using microneedles are as follows: increase drug efficacy, reduce body immune response, comfortable usability, convenient transportability, and no need to be administrated by experienced personnel [1].

### 6.2.2   Using 3D Shapes in Order to Target the Desired Tissues

Special 3D shapes can also be used to target specific tissues like spleen. As we know, due to the filtering nature of this tissue, only particular particles can be inserted into this tissue (nanoparticles with dimensions less than 200 nm), disk-shaped red blood cells (erythrocytes) with a diameter of 10 μm can also pass through this tissue. In other words, the transfiguration of conventional particle to a disk-shaped particle can increase the permeability of drugs in particular tissues [1].

### 6.2.3  Effect of the Particle Shape on the Specific Surface Area

One of the effects of changing particle shape is specific surface area variation. Consequently, the interface of drug particles with the surrounding environment is changed that affects the kinetics of the drug release. Specific surface area is proportional to the size of the particle. In other words, controlling the size of a particle can lead to determine the solubility and release profile of the drug [1].

### 6.2.4  Effect of the Particle Shape on Targeting Specific Cells

Surveys have illustrated that due to the higher surface area of nanorods in comparison to spherical particles, they interact better with cell's receptors that improve absorption of nanorod-shaped drugs. In addition, the particle shape changes from spherical shape to nanorod can result in longer stability of drug particles in the body circulation system since immune cells detect nanorods shape drug less than spherical ones. This phenomenon increases half-life time of the drugs and thus enhances the chances of successfully reaching targeted cells. Moreover, endothelial cell has more affinity with nanorod-shaped drug than spherical ones. On the whole, each particular cell accepts a particular shape, and the change in the particles shape can also increase the level of toxin secreted by cells [1].

Till now, many different methods have been introduced to manufacture different 3D shapes in nano- and microscale dimensions. These methods are able to specify the structure of drug particles and also carrier devices so that therapists can control the drug release rate in desired conditions. Among these methods, 3D printing is an emerging technology that enables researchers to fabricate various shapes in different dimensions.

## 6.3  3D Printing Technology

3D printing is a newly introduced technology in which a 3D structure is made by depositing layer-by-layer consecutively. Although this technology is originated from rapid prototyping manufacturing methods, which tries to manufacture cost-effective models quickly, nowadays this technology is considered as scalable fabrication process to manufacture complex designs which were difficult to be produced by traditional processes. Using 3D printing, technology in drug delivery applications has been approved by FDA for the 3D-printed orodispersible tablets (Spritam) that encouraged fabrication of other kinds and dosage forms [4, 5]. Nonetheless, for printing precision shapes with low tolerances, we still need to add some impurities such as UV-curing agents to biomaterials that can be used for accurate printing ink [1].

Various fabrication procedures such as stereolithographic (SLA), powder-based (PB) technology, selective laser sintering (SLS), fused deposition modeling (FDM), extrusion-based FDM technology, and semisolid extrusion (EXT) 3D printing are employed to deposit biomaterials layer by layer for manufacturing complex devices. For tailored functionality of a 3D printed device, a computer software is responsible to control the size, shape, and internal structure of the device. 3D-printed devices are able to store several drugs inside one tablet with accurate dosing of each one. Also, the structure, shape, and size of the devices determine the dissolution and diffusion rate of each drug [6].

## 6.3.1  Stereolithographic (SLA) 3D Printing

In stereolithographic (SLA) 3D printing, the 3D product is fabricated by curing (photo-polymerization) a photosensitive material such as low molecular weight polyacrylates, epoxy macromers or monomers via ultraviolet (UV) light or digital light projection (DLP) technique. A digital micro-mirror device specifies the locations supposed to be exposed to light. The light radiation triggers the chemical gelation reaction. The process is redone layer after layer until the complete structure of a model is built up. In addition to long running time of SLA 3D printing method, the other limitation of this technique is carcinogenesis of used resins. Their long-term stability is not also good enough due to their photosensitivity [7].

## 6.3.2  Powder-Based (PB) 3D Printing

In PB 3D printing, the powder is spread over a surface to make a thin layer by powder bed or powder jetting mechanisms, and then the specific area of the surface is applied by liquid binder drops that are ejected from an inkjet printer head. This technique has been applied to fabricate zero-order release as well as controlled-release tablets with complex release profiles such as immediate-extended or dual pulsatory [7]. PB 3D printing is an outstanding method especially for bone reconstruction, since all the bone properties such as porosity, shape, structure, and mineral properties can be imitated. The most used materials for bone reconstruction are calcium phosphate (CaP), inorganic bovine bone, hydroxyapatite (HA), and demineralized human bone matrix powders that are widely applied to 3D-printed scaffold bone. Recently, scientists have tried to incorporate growth factors and also drugs with osteoconductive scaffolds. Regarding the combination of growth factors with osteoconductive materials, the new bone formation has meaningfully enhanced; moreover, due to the tridimensional stability of the scaffold, the graft is protected while it is being substituted with newly formed bone [8].

### 6.3.3   Selective Laser Sintering (SLS) 3D Printing

SLS 3D printing is similar to PB 3D printing, but in SLS the layered powders are fused completely or partially by $CO_2$ or Nd:YAG laser radiations. The starting materials that can be used are classified into four types including polymers (polyamides, polystyrenes, polycarbonates, and nylon), ceramics (like HA, and also glasses and glass-ceramics), metals (steel, titanium, aluminum, and silver), and composites (HA/polyetheretherketone). Figure 6.3 illustrates the general steps of the SLS 3D printing process [9].

Although using SLS technology is well accepted in tissue engineering industry, it is not commonly used in pharmaceutical applications due to the feasibility of drug and pharmaceutical excipients' degradation by the beam energy [7].

### 6.3.4   Fused Deposition Modeling (FDM) 3D Printing

FDM or fused filament fabrication (FFF) is a cheap and prevalent process in which the 3D structure is formed by extruding the fully or partially melted layers from the printer's head at determined direction that is directed by printer software. In other words, the materials are heated inside the FDM instrument to the temperature higher than its softening point. After that, the fused material is extruded from the printer nozzle layer by layer and solidifies to create the 3D structure. In drug

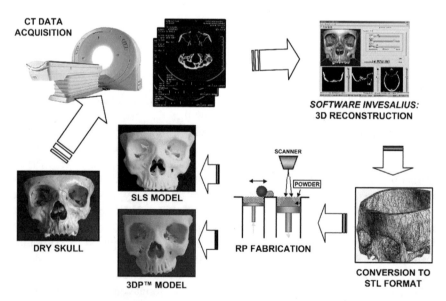

**Fig. 6.3** General process of SLS and 3DP™ model production (Reprinted from [10], Copyright (2008), with permission from Elsevier)

delivery applications, the drug is loaded on PVA by incubating it inside saturated drug solution. Till now different shapes of drug-loaded tablets such as pyramidal, cubic, cylindrical, spherical, and toroidal were successfully fabricated using FDM 3D printing. However, the high-temperature operation ($\sim 220$ degree centigrade) still limits the applicability of using this technique in pharmaceutical applications. In addition to that, because of the drug-loading process (incubation in saturated drug solution), low drug loading is inevitable in this technique [7].

## 6.3.5   Semisolid Extrusion (EXT) 3D Printing

EXT 3D printing, also known as the pressure-assisted microsyringe (PAM) printing method, is another method which is based on extrusion of semisolid materials (gels or pastes) layer by layer from a nozzle on a build plate. After depositing the first layer on the plate, it is lowered and the process is redone again and again until the structure become completed. This kind of 3D printing process was first applied to print living cells and other biologics for manufacturing living tissue and organs. So, it is mostly called extrusion-based bioprinting technique. It is worthy of mentioning that the term bioprint is used appropriately when living cells or other biologics are printed [5]. Although this process does not necessitate high temperatures for fabrication, it requires the materials to be in paste form, so it increases the shrinking and deformation failures of products during drying [7].

Overall, for manufacturing a product by 3D printing technology at first, we need a 3D model which is usually captured from computer-aided design and drafting (CADD), computed tomography scan (CT scan), or magnetic resonance imaging (MRI) data. Then, the prepared 3D model file is opened by the printer software.

**Table 6.1** Concept and minimum layer thickness (Preciseness) of 3D printing technologies [7]

| 3D printing tech. | Concept | Minimum layer thickness (μm) |
|---|---|---|
| SLA | UV radiating over the top of a photopolymerizable liquid in a layer-by-layer fashion | 100 |
| PB | The ink (binders and pharmaceutical ingredients) is poured over a powder bed in two-dimensional fashion to make the final product in a layer-by-layer fashion | 200 |
| SLS | A laser beam sinters the powder and binds it in layer by layer fashion | 100 |
| FDM | Extruding a thermoplastic filament through high-temperature nozzle into semisolid fused state filament in layer-by-layer fashion | 100 |
| EXT (PAM) | Extruding semisolids (e.g., homogeneous paste) over moveable stage in layer-by-layer fashion into a product | 800 |

The printer software cuts up the model into layers with specific height for each layer and determines how the layers are going to be printed and assembled layer by layer [5].

Table 6.1 briefly compares the concept and preciseness of each 3D printing technique [7].

# References

1. Curry EJ, Henoun AD, Miller AN, Nguyen TD (2017) 3D nano- and micro-patterning of biomaterials for controlled drug delivery. Ther Deliv 8:15–28. https://doi.org/10.4155/tde-2016-0052
2. Traverso G, Schoellhammer CM, Schroeder A, Maa R, Lauwers GY, Polat BE, Anderson DG, Blankschtein D, Langer R, Anderson D (2015) Microneedles for drug delivery via the gastrointestinal tract HHS public access. J Pharm Sci 104:362–367. https://doi.org/10.1002/jps.24182
3. Liu S, Wu D, Quan YS, Kamiyama F, Kusamori K, Katsumi H, Sakane T, Yamamoto A (2016) Improvement of transdermal delivery of exendin-4 using novel tip-loaded microneedle arrays fabricated from hyaluronic acid. Mol Pharm 13:272–279. https://doi.org/10.1021/acs.molpharmaceut.5b00765
4. Prasad LK, Smyth H (2016) 3D Printing technologies for drug delivery: a review. Drug Dev Ind Pharm 42:1019–1031. https://doi.org/10.3109/03639045.2015.1120743
5. Palo M, Holländer J, Suominen J, Yliruusi J, Sandler N (2017) 3D printed drug delivery devices: perspectives and technical challenges. Expert Rev Med Devices 14:685–696. https://doi.org/10.1080/17434440.2017.1363647
6. Zhang H, Jackson JK, Chiao M (2017) Microfabricated Drug Delivery Devices: Design, Fabrication, and Applications. Adv Funct Mater 1703606:1–31. https://doi.org/10.1002/adfm.201703606
7. Alhnan MA, Okwuosa TC, Sadia M, Wan K-W, Ahmed W, Arafat B (2016) Emergence of 3D printed dosage forms: opportunities and challenges. Pharm Res 33:1817–1832. https://doi.org/10.1007/s11095-016-1933-1
8. Brunello G, Sivolella S, Meneghello R, Ferroni L, Gardin C, Piattelli A, Zavan B, Bressan E (2016) Powder-based 3D printing for bone tissue engineering. Biotechnol Adv 34:740–753. https://doi.org/10.1016/j.biotechadv.2016.03.009
9. Shirazi SFS, Gharehkhani S, Mehrali M, Yarmand H, Metselaar HSC, Adib Kadri N, Osman NAA (2015) A review on powder-based additive manufacturing for tissue engineering: selective laser sintering and inkjet 3D printing. Sci Technol Adv Mater 16:33502. https://doi.org/10.1088/1468-6996/16/3/033502
10. Silva DN, Gerhardt de Oliveira M, Meurer E, Meurer MI, Lopes da Silva JV, Santa-Bárbara A (2008) Dimensional error in selective laser sintering and 3D-printing of models for craniomaxillary anatomy reconstruction. J Cranio-Maxillofacial Surg 36:443–449. https://doi.org/10.1016/j.jcms.2008.04.003

Printed in the United States
By Bookmasters